Intelligence in IoT-enabled Smart Cities

Intelligence in IoT-enabled Smart Cities

by
Fadi Al-Turjman

CRC Press
Taylor & Francis Group
Boca Raton London New York

CRC Press is an imprint of the
Taylor & Francis Group, an **informa** business

MATLAB® and Simulink® are trademarks of The MathWorks, Inc. and are used with permission. The MathWorks does not warrant the accuracy of the text or exercises in this book. This book's use or discussion of MATLAB® and Simulink® software or related products does not constitute endorsement or sponsorship by The MathWorks of a particular pedagogical approach or particular use of the MATLAB® and Simulink® software.

CRC Press
Taylor & Francis Group
6000 Broken Sound Parkway NW, Suite 300
Boca Raton, FL 33487-2742

Printed on acid-free paper
Printed by CPI Group (UK) Ltd, Croydon CR0 4YY

International Standard Book Number-13: 978-1-138-31684-3 (Hardback)

Library of Congress Cataloging-in-Publication Data

Names: Al-Turjman, Fadi, author.
Title: Intelligence in IoT-enabled smart cities / Fadi Al Turjman.
Description: Boca Raton : Taylor & Francis, a CRC title, part of the Taylor & Francis imprint, a member of the Taylor & Francis Group, the academic division of T&F Informa, plc, 2019. | Includes bibliographical references.
Identifiers: LCCN 2018045099| ISBN 9781138316843 (hardback : acid-free paper) | ISBN 9780429022456 (ebook)
Subjects: LCSH: Smart cities. | Internet of things. | Municipal engineering--Technological innovations.
Classification: LCC TD159.4 .A48 2019 | DDC 307.1/160285--dc23
LC record available at https://lccn.loc.gov/2018045099

Visit the Taylor & Francis Web site at
http://www.taylorandfrancis.com

and the CRC Press Web site at
http://www.crcpress.com

To my dearest parents, my brother, and my sisters.
To my wonderful wife and my little prince.

Contents

Preface

As one of the most significant topics in IoT-based systems, the Smart Cities and Smart Sensory platforms have emerged to intelligently address major challenges in our daily life. These challenges can vary from cost and energy efficiency to availability and service quality. It is our aim to focus on both design and implementation aspects in the application of smart cities that are enabled by intelligent wireless sensor networks and other enabling technologies in the IoT era. Wireless sensor networks are a key enabling technology and should be smart enough to offer multiple sustainable and viable benefits from emerging smart city paradigms and its interacting technologies such as radio frequency identification, wireless sensor networks (WSNs), LTE-A, and 5G. This book provides a comprehensive overview for the use of sensory platforms in intelligent/smart IoT environments in line with the smart city paradigm. These platforms integrated with intelligent algorithms can help mobile operators and service providers afford the next generation of this paradigm. In this book, we overview the interactions and data access of smart devices and provide a review of their applications in smart IoT-based systems.

Fadi Al-Turjman

MATLAB® is a registered trademark of The MathWorks, Inc. For product information, please contact:

The MathWorks, Inc.
3 Apple Hill Drive
Natick, MA 01760-2098 USA
Tel: 508 647 7000
Fax: 508-647-7001
E-mail: info@mathworks.com
Web: www.mathworks.com

Author

 **Prof. Fadi Al-Turjman** received his Ph.D. degree in computer science from Queen's University, Canada, in 2011. He is a Professor with Antalya Bilim University, Turkey. He is a leading authority in the areas of smart/cognitive, wireless and mobile networks' architectures, protocols, deployments, and performance evaluation. His record spans over 180 publications in journals, conferences, patents, books, and book chapters, in addition to numerous keynotes and plenary talks at flagship venues. He has authored/edited more than 12 published books about cognition, security, and wireless sensor networks' deployments in smart environments with Taylor & Francis, and the Springer (Top tier publishers in the area). He was a recipient of several recognitions and best papers' awards at top international conferences. He led a number of international symposia and workshops in flag-ship ComSoc conferences. He is serving as the Lead Guest Editor in several journals, including the IET Wireless Sensor Systems and Sensors (MDPI and Wiley). He is also the Publication Chair for the IEEE International Conference on Local Computer Networks.

Chapter 1

Introduction

Fadi Al-Turjman

Contents

The population of the world is increasing rapidly. It was 7.6 billion in October 2017. Currently over 200 million people are living in urban cities and this number is expected to increase up to 5 billion by 2030. Megacities are also increasing in terms of size and number, and this brings various problems with it. Environmental deprivation, such as the uncontrolled noise, waste pollution, low management of consumption of the non-renewable energy resources, deficits in water supply and waste collection became a very serious problem [1]. Creating the infrastructure of smart cities is considered a promising future goal for creating a sustainable environment, where the main components of these cities, for example, buildings, roads, energy stations, water pools, etc., are all managed by Internet of Things (IoT) devices which sense, collect, and transmit the necessary data among them [2]. In other words, a smart city is an urban area which utilizes various types of electronic data collection sensors to supply information used to control and manage assets and resources efficiently. In fact, a *smart city* has no exact definition. According to Hall R.E. a smart city is "a city that monitors, and integrates conditions of all its critical infrastructures, including roads, bridges, tunnels, rails, subways, airports, seaports, communications, water, power, even major buildings can better optimize its resources, plan its preventive maintenance activities, and monitor security aspects while maximizing services to its citizens" [3]. Other definitions found in the literature are listed in Table 1.1.

Table 1.1   Definitions of Smart City

Reference	Definition
[4]	"The use of smart computing technologies to make critical infrastructure components and services of a city—which include city administration, education, healthcare, public safety, real estate, transportation, and utilities—more intelligent, interconnected, and efficient."
[5]	"A city well-performing in a forward-looking way in economy, people, governance, mobility, environment, and living, built on the smart combination of endowments and activities of self-decisive, independent and aware citizens."
[6]	"A city that gives inspiration, shares cultures, knowledge, and life, a city that motivates its inhabitants to create and flourish in their own lives."
[7]	"A city where the ICT strengthen the freedom of speech and the accessibility to public information and services."

Nowadays, there are lots of smart city projects in developing and developed countries. Table 1.2 contains a cumulative list of cities that earned high scores from the Intelligent Community Forum (ICF) from 2007 to 2011 by accomplishing factors of intelligent communities successfully. These factors are as follows: innovation, marketing and advocacy, digital inclusion, knowledge, and broadband connectivity [8].

Numerous studies have been done in the field of components of smart cities and infrastructure of smart cities. Amirhosein et al. [9] examined the different notions of smart buildings while analyzing from an international perspective. As a conclusion, the author claims that, smart buildings play a fundamental role in shaping future cities. He supports the infrastructure of smart buildings because of their reduced environmental impacts, low operational cost, improved security systems, etc. In another study by Ibrahim Dinçer et al. [10], the author describes smart energy systems for a sustainable future by evaluating each of the energy sources by their efficiencies, environmental performances, and energy and material sources. He concludes that, if products from the same energy source are high, greater efficiency can be achieved with lower emission levels. He also found that, geothermal energy is a cleaner and more efficient resource for the future. In Robles et al., the authors highlights an Internet of Things (IoT)-based model for smart water management, monitored from the business perspective. The model he proposed can be implemented not only in urban cities but also in rural areas, and agricultural applications. In conclusion, the authors suggest that the development of IoT with the integration of their models in smart cities can be feasible, scalable, and industrial. In Snellen and Hollander, the authors have emphasized the role of smart mobility and transportation as a result of the development of Information and Communication Technologies (ICT) in smart cites as well. He reviewed the effects

Table 1.2   List of Smart Cities in Different Regions Worldwide

Region	Cities
Asia	Shanghai, Tianjin (China); Doha (Qatar); Gangnam District, Seoul, Hwaseong-Dongtan, Sowen (Korea); Hyderabad, Bangalore, Jaipur, Rajasthan (India); Ichikawa, Mitaka, Yokosuka, Jia Ding (Japan); Kabul (Afghanistan); Hong Kong; Taipei, Taoyuan Country (Taiwan); Tel-Aviv (Israel)
Africa	Cape Town, Nelson Mandela Bay (South Africa)
Europe	Besancon, Issy-lex-Moulineaux (France); Birmingham, Dundee, Scotland, Glasgow, Manchester, Sunderland (UK); Eindhoven (Netherlands); Hammarby Sjostad, Karlskrona, Stockholm (Sweden); Malta (Malta); Reykjavik (Iceland); Sopron (Hungary); Tallinn (Estonia); Trikala (Greece)
North America	US; Albany, Westchester Country (New York); Ashland (Oregon); Arlington Country, Danville, Bristol (Virginia); Bettendorf (Iowa); Chattanooga (Tennessee); Cleveland, Dublin, Northeast Ohio (Ohio); Corpus Christi (Texas); Dakota Country (Minnesota); Florida High-Tech Corridor (Florida), LaGrange (Georgia); Loma Linda (California); San Francisco; Spokane (Washington); Winston-Salem (North Carolina); Canada; Burlington, Ottawa, Kemora, Strafford, Toronto, Waterloo, Windsor-Essex (Ontario); Moncton (New Brunswick); Quebec City (Quebec); Winnipeg (Manitoba); Vancouver (British Columbia); Western Valley (Nova Scotia); Calgary (Alberta)
Middle/South America	Barceloneta (Puerto Rico); Curitiba, Parana, Pieria, Porto Alleger (Brazil)
Oceania	Balart, Gold Coast City, Ipswich, Queensland, Victoria, Whittlesea (Australia)

of technological changes to our society, the public, and our traditions. People should know what they want and what they need and start creating new frameworks in order to bridge the policy gap."

In this book, we overview the advantages of intelligent algorithms and approaches in smart cities toward further improving the quality of our daily life while highlighting key design challenges. Accordingly, our main contributions in this work can be summarized as follows.

■ We start by overviewing the importance of energy consumption methods in smart cities and their monitoring techniques. We highlight possible strategies that can be applied for further optimization in energy savings.

- Various techniques and tools available for smart city applications such as smart homes, smart grid, smart parking, smart agriculture, and smart vehicular networks in IoT environments are also presented that have the potential to realize more intelligent energy-saving solutions. These solutions and energy-related challenges are discussed in detail.
- We describe prominent performance metrics in order to understand how the energy efficiency is evaluated and how data is better exchanged.
- Finally, we conclude this book with open security and privacy issues in mobile applications in the smart city paradigm.

Book Outline

The rest of this book is organized as follows. In Chapter 2, we delve into an overview of energy monitoring in smart cities. In Chapter 3, we describe prominent performance metrics in smart home applications and communication protocols. Chapters 4 and 5 provide energy-based analysis for the smart grid applications in smart cities. Chapter 6 proposes an overview of the smart parking projects while highlighting technical issues and design factors in their communication protocols. Chapter 7 emphasizes the intelligent medium access control in smart city transportation systems. In Chapter 8, we propose a multimedia-inspired approach for data delivery while tolerating failure in smart city's mobile applications and satisfying Quality of Service (QoS) parameters. In Chapters 9 and 10, we focus on cognitive data delivery and positioning approaches in the smart city agriculture paradigm. In Chapter 11, we investigate security and privacy issues in IoT-related smart city enabling technologies such as the Vehicular Ad-hoc Networks (VANETs).

References

1. F. Al-Turjman, "5G-enabled Devices and Smart-Spaces in Social-IoT: An Overview", *Elsevier Future Generation Computer Systems*, 2017. DOI: 10.1016/j.future.2017.11.035.
2. R. Arshad, S. Zahoor, M. Ali Shah, A. Wahid, and H. Yu. "Green IOT: An Investigation on Energy Saving Practices for 2020 and Beyond," *IEEE*, vol. 5, pp. 15667–15681, 2017.
3. F. Al-Turjman. "Mobile Couriers' Selection for the Smart-grid in Smart cities' Pervasive Sensing", *Elsevier Future Generation Computer Systems*, vol. 82, no. 1, pp. 327–341, 2018.
4. F. Al-Turjman and S. Alturjman. "5G/IoT-Enabled UAVs for Multimedia Delivery in Industry-oriented Applications", *Springer's Multimedia Tools and Applications Journal*, 2018. DOI. 10.1007/s11042-018-6288-7.
5. R. Giffinger, C. Fertner, H. Kramar, R. Kalasek, and N. Pichler-Milanovic, "*Smart Cities: Ranking of European Medium-Sized Cities*," Centre of Regional Science, Vienna, 2007.

6. F. Al-Turjman, and S. Alturjman, "Context-sensitive Access in Industrial Internet of Things (IIoT) Healthcare Applications", *IEEE Transactions on Industrial Informatics*, vol. 14, no. 6, pp. 2736–2744, 2018.
7. H. Partridge, "Developing a human perspective to the digital divide in the smart city," in *ALIA 2004 Challenging Ideas*, Queensland, 2004.
8. T. Nam and T.A. Pardo. "Conceptualizing Smart City with Dimensions of Technology, People, and Institutions," in *12th Annual International Conference on Digital Government Research*, Maryland, 2011.
9. F. Al-Turjman, "The Road Towards Plant Phenotyping via WSNs: An Overview", *Elsevier Computers & Electronics in Agriculture*, 2018. DOI: 10.1016/j.compag.2018.09.018.
10. C. A. Ibrahim Dincer, "Smatr Energy Systems for a Sustainable Future," *Applied Energy*, vol. 194. 225–235, 2017.
11. T. Robles, R. Alcarria, D., Martin, M. Navarro, R. Calero, S. Iglesias, and M. Lopez. "An Internet of Things-based model for smart water management," in *28th International Conference on Advanced Information Networking and Applications Workshops*, Victoria, Canada, 2014.
12. D. Snellen and G. de Hollander. "ICT'S change transport and mobility: mind the policy gap!," in *44th European Transport Conference 2016*, Barcelona, Spain, 2016.

Chapter 2

Energy Consumption Monitoring in IoT-based Smart Cities

Fadi Al-Turjman* and Chadi Altrjman[†]

Contents

* Antalya Bilim University, Antalya, Turkey
† University of Waterloo, ON, Canada

2.1 Introduction

The Internet of Things (IoT) has been recently recognized as a disruptive technology for the flying ad hoc network. IoT can be viewed as a network of networks. There can be a wide range of applications in IoT that supports logistics and the management of flying ad hoc networks. IoT technology can be leveraged to achieve cost reductions. IoT technology can be combined with real-time location systems to get live updates from the factory floor, enabling manufacturers to continuously monitor machine activity, maintenance needs, and also product movement during production. Cost reduction can be achieved across the digital ad hoc networks by making use of these smart machines by providing data that allows manufacturers to adjust production on the fly. Manufacturing and assembly lines will receive updated schedules and quality-related information in real time and instantly. IoT data can be leveraged to schedule, maintenance, customized production to meet customers' orders, proactive, preventative, and predictive repairs, and to sharpen the focus that is needed to be successful in the digital world. The concept of Industry 4.0 aims at achieving a smart factory will soon be a reality. Smart products which consist of the embedded knowledge of their customers' needs will provide data insights and analytics about the best way to achieve customer fulfillment. All this information will lead to more cost-efficient production and product development.

  The use of IoT in ad hoc networks management is increasing at a rapid pace. Deutsche Bahn, the German Railways and cargo carrier, installed a network-wide monitoring system to manage its entire rail network which comprises over 1 billion supply chain "nodes", collecting data on each segment of track, rail car, station, engine, and switch, and monitoring the condition of all of these things in real time. The collected data are fed into a control tower that aggregates them every five seconds to provide near-real-time information across the entire fleet.

Deutsche Bahn has used these data to improve risk management practices such as real-time rerouting and optimization, considering all existing network traffic through nodes. Whirlpool is another example of using the IoT for internal supply chain optimization in routing work and locating misplaced inventory. Instead of using bar codes or a similar solution, Whirlpool used radio frequency identification (RFID) tags and readers across a manufacturing plant to give managers and operators real-time access to information for inbound logistics to the paint line.

The IoT can be used by the various partners of the supply chain to monitor its execution process in real time and improve the efficiency and effectiveness of the energy consumption. Recent developments in the IoT have made it possible to achieve high visibility in the supply chain. The location of arbitrary individual things can be determined at any point in time by all appropriate supply chain partners. For example, the IoT benefits the food and agricultural product supply chain by improving the visualization and traceability of agriculture products and ensuring people's food safety. Industrial deployment of the IoT provides development of an ideal platform for decentralized management of warehouses and collaborative warehouse order fulfillment with RFID, ambient intelligence, and multiagent systems.

Notwithstanding the huge enthusiasm by supply chain managers to use the IoT, there is still a lack in energy consumption monitoring and optimization. Observing and monitoring frameworks in this domain can gather and send key information about the monitored resource condition, hardware execution, testing, vitality utilization, natural conditions, and enable administrators and mechanized controllers to react to changes progressively anywhere. These capacities are vital to key production network exercises where deceivability and traceability of articles are required.

RFID is a piece of the IoT frameworks identified with inventory network exercises. RFID permits programmed recognizable proof and information catch utilizing radio waves, a tag, and a per-user. The tag can store more information than customary scanner tags. The tag contains information such as the Electronic Product Code (EPC), a worldwide RFID-based thing distinguishing proof framework created by the Auto-ID Center. RFID per user can recognize, track, and screen the items connected with labels all around, naturally, and progressively, if necessary.

Notwithstanding the RFID, remote advancements have assumed a key part in mechanical observing and control frameworks. In an average production network, observing and control applications utilize sensors, GPS, RFID labels, and sensor systems to limit scattering, robbery, misfortune, and decay in distribution center, transportation, and store racks [33]. Sensors are utilized to keep merchandise at the right temperature and shield them from substance spills that lead to waste [34]. Sensor systems screen movement conditions, and route gadgets track the area of transportation vehicles to make steering more proficient. The Wireless Sensor Network (WSN) comprises spatially dispersed self-sufficient sensor prepared gadgets to screen physical or ecological conditions and can participate with RFID frameworks to better track the status of objects, for example, their area, temperature, and developments [35].

WSN has been principally utilized as a part of frosty chain coordinations that includes the transportation of temperature-sensitive items along an inventory network through warm and refrigerated bundling strategies [36]. WSN is additionally utilized for upkeep and for following frameworks. General Electric (GE) uses sensors in its fly motors, turbines, and wind ranches. By dissecting information continuously, GE spares time and expenses because of convenient preventive support. American Airlines additionally utilizes sensors fit for catching 30 terabytes of information for every flight for administrations, for example, safeguard upkeep.

The communication between the utilities and the consumer in supply chain helps a lot in terms of energy conservation. The behaviors of the consumer vary not only with the changes in tariff but is also affected if the consumer is aware of appliance's consumption. This direct feedback helps in conserving energy as well as managing the resources for both customer and supplier. In order to understand the importance of monitoring, let's take an example of a gas station, where all members from one family go to refill their car's fuel tanks but no one pays after refilling because they have a monthly billing system with the gas station owner, where they pay a lump sum bill at the end of month. At the end of the month when the family gets the bill, they don't know who they must account for such a large bill from the gas station.

There, our electricity consumption is the same; to simplify it swap the gas station in the example with the electricity supplier company. Receiving the bill at the end of the month and not knowing which appliance is contributing how much to the bill doesn't help to conserve energy. This is the part where energy monitoring helps provide a better solution in order to utilize power efficiently while aiding conservation by monitoring where the energy is being used and what can be done to reduce the consumption.

Smart Energy Monitoring (Direct Feedback) works on a very basic principle where the energy being supplied by the provider company enters the home by Watt Hour Meter (WHM) from where it goes to Load Survey Meter (LSM), which calculates the total load of the entire building. Further down in the system are End User Meters (EUMs) installed separately for all the appliances which calculates the energy consumption for individual appliances. At the end all the data from the ESMs is collected and distributed to a monitoring server.

Depending on the system the user has, the distribution server can be omitted from the system. If the user is using In-Home Monitoring Display (IHD) then the data from all the ESMs will be collected and processed to be illustrated on the IHD screen, but if the user is connected to a monitoring system provided by the supplying company then the data will be distributed to the supplier, conditioned, and then later the user will be informed about the consumption and given conservation tips (see Figure 2.1).

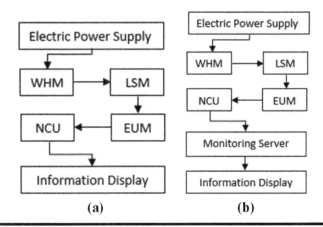

Figure 2.1 **Smart energy monitoring scheme; (a) In-house monitoring system, and (b) supplier controlled monitoring system, where WHM is** *Watt Hour Meter,* **LSM is** *Load Survey Meter,* **EUM is the** *End User Meter,* **and NCU is the** *Network Control Unit.*

The informational terminal, regardless of the type of monitoring system the user has installed, comprises a screen which has a very detailed logging of the energy consumption of the entire building. The display illustrates the electricity charges, daily load curves, daily changes in consumption over the last few days or week, and a comparison between past data for both individual appliances as well as the entire building, along with energy conservation tips as shown in Figure 2.2.

Energy conservation is a big challenge for both the customer and the supplier, but the beneficiaries of conservation are again, both. The consumers who participate in energy conservation gets the benefit of a direct decrease in their electricity bills as

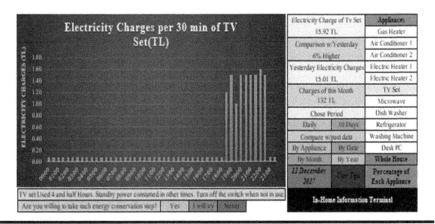

Figure 2.2 **In-home information display terminal.**

well as a lower carbon footprint. On the other hand, at the supplier level, conservation of energy means a continuous monitoring on consumption growth which may yield economic benefits by saving the expenditures on extra capacity investments as well as peak shifting helping to ensure continuous supply. In addition, it may also help the country by lowering the GHG emissions and relying on renewable resources.

A detailed review study of different pilot projects and responses is conducted to compare different monitoring systems, make conclusions from the results, and to suggest which is the better monitoring system. Also, the impact and effectiveness of general energy monitoring is discussed. Section 2.2 of the chapter categorizes and discusses the energy monitoring systems based on feedback system as well as subcategorizing them based on type of pricing policies. Section 2.3 briefly lists the devices used for sensing in a monitoring system. Section 2.4 discusses in detail the background research in a similar field and the results of pilot projects in different regions of the world. Sections 2.5 and 2.6 contain the conclusion of the results and the discussion, respectively.

2.2 Categories of Energy Monitoring System

Considering the energy monitoring system, there is considerable variety based on type of feedback system being used. Feedback is the most essential and helpful tool in learning format for the user, allowing them to learn through experimentation and controlling of the consumption. The categories of energy monitoring based on type of feedback system are discussed in the following sections.

2.2.1 Indirect Feedback Systems

An indirect feedback system is controlled by the utility (power supplier companies) in which the raw data collected by the utility is processed and later sent to the customer with the electricity bills. The drawback of this kind of system is the lack of real-time monitoring, which means the consumer is not helped to save the energy efficiently.

2.2.2 Direct Feedback Systems

A direct feedback system performs a real-time energy monitoring with the help of energy meters and IHD to visualize the power consumption information in addition with electricity rate and usage. The data from every appliance is collected and processed to form a consumption trend. Direct feedback systems are subcategorized based on the pricing policy being used which is discussed in the next section.

2.2.2.1 In-Home Display Only

In this type of direct feedback system, the users are provided with an IHD monitor. Users with IHD have a monthly billing system in which price for consumption is time independent and static for an entire day.

2.2.2.2 In-Home Display with Pay-As-You-Go

Pay-As-You-Go is a pricing policy in which the users of electricity are entitled to buy the electricity prior to usage with help of a Smart Card which can be recharged from Utility Customer Centers. Later, the user plugs in the Smart Card which activates the energy meter. As soon as the card runs out of prepaid energy, the meter cuts the electricity connection.

2.2.2.3 In-Home Display With Time of Use Pricing Policy

Time of Use (TOU) is a pricing policy in which the price per unit consumption of electricity is time dependent i.e., the price of electricity consumption per unit is higher in peak hours of the day and lower in off-peak hours. The users of TOU are aware of per unit price during peak and off-peak hours. In addition to this IHD monitors are also provided which differentiate the consumption trends, during peak and off-peak hours, by using color schemes.

2.3 Energy Monitoring Devices

The fundamental components of the energy monitoring are the sensing devices, to be used to measure or sense the energy consumption for individual devices. In order to calculate the power consumed by every appliance two basic variables are to be known i.e., voltage and current.

In order to measure these variables (current and voltage) various methods using different devices are used which are discussed as follows:

2.3.1 Direct Sensing

Direct sensing is the most basic method used in the electrical field, in which the current and the voltage of the appliance can be measured directly by using a voltmeter and an ammeter. In this method the current in the appliance is measured in series using an ammeter whereas the voltage across the appliance is measured in parallel using a voltmeter. To avoid the usage of two different meters a multi-meter can be used to measure both current and voltage. Since these variables are to be measured across every appliance, the application of a multi-meter is unfeasible in terms of financial optimization. In order to overcome this problem, a microprocessor can be programmed (along with sensing elements for voltage and current), which is the more economical option for measuring the variables.

2.3.2 Indirect Sensing

An indirect power monitoring sensing system is one of the promising systems being used in household monitoring systems, in which an energy interference engine is built using magnetic, light, or acoustic etc. sensors to indirectly sense the signals emitted from the appliance in order to measure the current and voltage to calculate power consumption. The system involves radio enabled sensors distributed in entire house which sense the signals emitted from appliances and forward it to a computing system, which collects data from all sensing elements and the main power meter, later running a pre-saved algorithm to estimate the power consumption of each appliance. The implementation uses Crossbow MicaZ wireless sensor nodes to run magnetic, light, and acoustic sensors. The sensors most used for indirect sensing are discussed in detail later.

2.3.2.1 Magnetic Sensors

HMC 1002 magnetic sensors are surface mounted sensors which are designed for low field magnetic sensing as well as navigation system, magnetometer, and current sensing by the use of Anisotropic Magneto-Resistive (AMR) technology which gives it a leverage over coil-based sensors. This magnetic sensor measures the magnetic field changes near the power cables for appliances, as the magnetic field variation is strongly related to current flow in the wires thus making it good for current sensing [2].

2.3.2.2 Light and Acoustic Sensors

A number of possible options are available for the placement of light and acoustic sensors in a monitoring system. The most commonly used sensors are CdSe photocell light intensity sensors and simple piezoelectric microphones [1]. These light and acoustic sensors are used to sense the internal power stats of an appliance in terms of noise pattern and light intensity.

As mentioned, a vast number of options are available for remotely sensing for power consumption estimation, depending on the feasibility and calibration according to environment. The most appropriate sensor must be chosen to acquire the best estimated results

2.4 Literature Review

A number of studies have been conducted to evaluate the impact and effectiveness of a feedback system on consumer's behavior, which have consistently suggested that consumer's response to feedback system is always positive in terms of energy conservation and most of the studies suggest a decrease ranging up to 20 percent. The geographical locations of the pilot programs discussed as well as the demographics of the consumer vary from place to place according to societal regimes and weather conditions. A number of pilot projects will be discussed, in which the results will be later compiled and compared based on the type of project.

2.4.1 Pilot Projects with IHD Only

This part of the review considers different pilot projects in Canada, the United States, and Japan that used IHD to observe the impact on consumer's behavior.

2.4.1.1 Hydro-One Project, Canada

The Hydro One Networks in the state of Ontario, Canada started a pilot project in 2004 in which around 400 participants were included from the city of Barrie, Brampton, Lincoln, Peterborough, and Timmins. The consumption patterns of the participants were monitored over a period of two and half years. Later the participants were provided with an IHD for power cost monitoring, without any rate incentives, to monitor their consumption trends and the impact of IHD was observed by the suppliers. The real-time feedback provided by the power cost monitoring in-home display yielded an average decrease of 6.5 percent among all the participants. At the end of the pilot project, 65.1 percent of the users continued to use the power consumption monitor IHD for energy conservation.

2.4.1.2 CEATI Pilot Project, Canada

Similar to the Hydro One project, the Customer Energy Solution Interest Group (CEATI) comprising of BC Hydro, Newfoundland Power, National Rural Electric Cooperative Association and Natural Resource Canada Office of Energy Efficiency undertook a project in the state of British Columbia, Newfoundland and Labrador where a group of around 200 users of Newfoundland Power and BC Hydro. Consumers were convinced to participate provided that no rate incentives would be given. Their trends were monitored for a period of one and a half years and later were provided with a power cost monitor as an in-home display. The preliminary results of the project found an almost 18 percent average decrease in consumption of participants from Newfoundland Power and 2.7 percent decrease in consumption of BC Hydro users.

2.4.1.3 Cost Monitoring Pilot Project, US

In the year 2007, three major suppliers, in the state of Massachusetts in the United States, namely National Grid, NSTAR, and Western Massachusetts Electric

Company conducted a similar pilot project as mentioned in earlier project sections by distributing a number of 3,512 IHDs in which NSTAR had the major contribution. The project mainly focused on cost benefit analysis and customer opinion on savings, so only the estimated results were published as half of the customers perceived around 5–10 percent savings in their billings.

The pilot project was concluded to be effective in terms of energy conservation as 63 percent of the total participants reported that they have changed their behavior toward electricity consumption just by cutting down the standby consumption. Among them 41 percent of users reported to be more conscious toward switching off lights when not in use, 23 percent for TV, 18 percent for computer, and 17 percent for battery charger use. The users reported that they managed to reduce 5–10 percent of their electricity bills by just cutting off supply for standby energy consuming appliances.

2.4.1.4 SDG&E Pilot Project, US

In 2007, San Diego in the state of California in the United States started a pilot project with a number of 300 participants. IHDs were installed for the residents whose monthly consumption was more than 700 kWh. The difference in this project compared to the one mentioned before was the phone calls and email tips and notification for conservation and consumption trends. The resulting data demonstrated a decrease of 13 percent of energy consumption compared to the previous year. The decrease in energy consumption was reported to be the result of both the IHD as well as the phone calls and emails of conservation tips.

2.4.1.5 ECOIS Project, Japan

In February 2000, Osaka University of Japan commenced a pilot project using an energy consumption information system with nine households to monitor their energy over a time period of two (2) years and later in January 2002 an IHD was installed.

The ECOIS display used was very detailed and customized for better options and feedback tips. The system was monitored by the servers for providing conservation tips. Results suggested a decrease of 9 percent in consumption after the ECOIS installation. Figure 2.3 represents the trends of energy consumption before and after the installation of ECOIS where it can be seen that energy consumption was reduced by a significant amount after consumers started to monitor their consumption.

2.4.2 Pilot Projects with IHD and Prepay Method

The second category in the field of energy monitoring is where the customers are engaged to a pilot program using IHDs but users are obliged to purchase the power

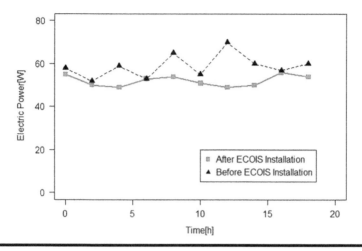

Figure 2.3 **Load curve before and after installation of IHD [6].**

prior to the usage. As mentioned earlier, smart cards are used to purchase electricity and plugged inside an IHD or an energy meter to activate electricity usage. Pilot projects following this method are discussed as follows.

2.4.2.1 SRP "Pay-As-You-Go" M-Power Project, US

In the state of Arizona, a project named "Salt River" was commenced, where 100 participants were provided with IHD as well as prepay meters by using smart cards to purchase electricity.

An average of 12.8 percent reduction was observed in energy consumption (13.8 percent in summer and 11.1 percent in winter). The M-Project is still active as one of the biggest programs in Northern America with a number of more than 50,000 subscribers [8, 9].

2.4.2.2 Woodstock Hydro's "Pay-As-You-Go," Canada

In the state of Ontario, Canada, a city named Woodstock launched a program in which 2,500 consumers voluntarily agreed to be a part of a pilot program in 2004. The number of consumers represented around one-quarter of the Woodstock Hydro's customers. The customers who took part in this pilot project were entitled to a monthly fee for administration and one-time installation cost. Besides that, the consumers were using a smart card for a prepay system in order to purchase the power before using it (commonly known as the Pay-As-You-Go method), later installing the card to an IHD resulting in activation of an attached electronic meter to provide electric power which was quite similar to the SRP M-Power Project.

As the results were compiled and processed, an average decrease of 15 percent was observed in the energy consumption of the consumers in the pilot project which is almost consistent with the results observed in the SRP M-Power Project with a decrease of 12.8 percent. Similarly, the results compiled include the effect of both IHD and Pay-As-You-Go, the effect of IHD and Prepay Power System were not identified individually.

2.4.3 Pilot Projects with IHD and Time Varying Rates

This is third category in the field of smart energy monitoring usually known as TOU in which the users are entitled to a time varying rate according to the time of power being used. As being supported through the chapter that energy monitoring has a high impact on user's behavior, equally is the time varying tariff effective in terms of energy conservation. Usually the rates are fixed higher at the peak hours to reduce the demand and set at lower price for off-peak hours. Often incentives are also given to promote less consumption during peak hours. Following are some case studies conducted in the category of IHD installed along with TOU price policy.

2.4.3.1 Hydro One TOU Project, Canada

Hydro One Networks in the state of Ontario, Canada started a pilot project in the summer of 2007 with 486 participants implementing TOU pricing policy along with IHD. Among all participants, 153 were provided with an IHD as well as TOU pricing, whereas 177 customers were just entitled to TOU pricing policy but were not provided with any IHD and an incentive of 50 USD at the end of program. Eighty-one participants of the project received just an IHD installation but no pricing policy or incentive. The rest of the project participants (75 users) were included as a control group.

The group with both IHD and TOU resulted to conserve 7.60% energy whereas group with TOU only saved 3.30 percent. The group using IHD without TOU managed to conserve 6.70% energy suggesting; IHD alone was more effective then TOU alone, moreover, when the pricing policy was introduced along with IHD, the energy conservation also raised a significant amount.

2.4.3.2 IDP Project, US

The Information Display project in the state of California in the United States had a user number of 61 among which 32 were residential users and the rest were commercial users. The user group was provided with IHD with TOU pricing policy.

The numerical results of the project were not evaluated as the program objective was only to observe the consumer demand response effect, but a post-project survey was conducted for residential and commercial users taking part in the project.

70 percent of residential and 65 percent of commercial users stated that their electricity consumption behavior was changed after the installation of a monitoring system, which led to energy conservation.

2.4.3.3 Home Energy Efficiency Trial, Australia

In 2004, Country Energy of Australia piloted a project named "Home Energy Efficiency Trial" for a period of 18 months, including 200 households which were provided with IHD and TOU.

A median saving of 8 percent was reported in the results and the customers claimed a reduction in electricity bills of around 16 percent, triggered by the behavior after installation of a monitoring system with TOU pricing policy. Additionally, 30 percent of demand decrease was noticed in the peak hours due to pricing policy.

Other major studies have also been conducted to determine the effectiveness of smart monitoring systems. The results for the conservation of energy derived from these pilot projects have been compared and categorized according to the type of feedback system in Table 2.1.

2.5 Results and Discussions

The literature suggests that the real-time monitoring yields a high percentage of energy conservation, but when pricing policy is introduced with energy monitoring, the results suggest that there is extra conservation. If the consumers interact with the IHD effectively, they can reduce their energy consumption by an average of 7 percent (using only the IHD method), whereas when using an IHD along with the prepayment method, energy conservation increases average up to 14 percent [14]. If compared, energy monitoring along with TOU pricing policy turns out to be better than others with the significant percentage of energy conservation. Similar studies as discussed in Section 2.5 have been reviewed and categorical comparison between them is illustrated in Figure 2.4.

2.6 Conclusion

A number of research studies and pilot projects have been conducted across the world, all suggesting the same result i.e., the energy monitoring system has been a very beneficial system in terms of energy conservation as the real-time feedback to the consumer induces a behavior change around consumption and conservation. [14]. Different scenarios have been discussed in the chapter to compare and contrast the results to evaluate the impact and effectiveness of smart energy monitoring.

Table 2.1 Energy Conservation Results of Various Feedback Pilot Projects

	Study	Duration	Sample Size	Location	Energy Conservation
Indirect feedback system	Bittle, Valesano, and Thaler [16]	2 months	353	US	8%
	Seligman et al. [17]	3 weeks	15	US	10%
		4 weeks	80	US	13%
	Arvola et al. [18]	2 years	525	Finland	3%
	Haakana et al. [19]	2.5 years	105	Finland	7%
		2.5 years	79	Finland	4–5%
	Wilhite and Ling [20]	3 years	191	Norway	10%
	Wilhite [21]	1 year	2000	Norway	4%
	Henryson et al. [22]	Unknown	600–1500	Scandinavia	2%
					3%
					2–4%
Direct feedback system	Seligman et al. [17]	4 weeks	10	US	16%
	McClelland and Cook [23]	11 months	25	US	12%
	Gaskell et al. [24]	4 weeks	80	UK	9%
	Winett et al. [25]	3 weeks	85	US	15%
	Hutton et al. [26]	Unknown	75	US, Canada	7%
	Sluce and Tong [27]	5 months	31	UK	13%
	Tsuyoshi Ueno et al. [6]	2 years	9	Japan	9%
	Van Houwelingen et al. [28]	1 year	50	Netherland	12%
	Dobson and Griffin [15]	2 months	25	Canada	13%
	Brandon and Lewis [29]	9 months	120	UK	12%
	NIE [15]	1 year	26	UK	4%
	Mountain et al. [30]	2.5 years	382	Canada	6.5%
	Benders et al. [31]	5 months	137	Netherland	8.5%
	Hydro One Network [3]	2.5 years	400	Canada	6.5%
	CEATI [4]	1.5 years	200	Canada	2.7%
	Green, A. [5]	1 years	300	US	13%

(Continued)

Table 2.1 (CONTINUED) Energy Conservation Results of Various Feedback Pilot Projects

	Study	*Duration*	*Sample Size*	*Location*	*Energy Conservation*
Direct feedback+ pay as you go	NIE [15]	12 months	35	UK	11%
	Pruitt B. [7]	Unknown	100	US	12.8%
	Quesnella K. [10]	Unknown	2500	Canada	15%
Direct feedback + time of use[a]	Sexton et al. [15]	Unknown	480	US	26%
	Owen and Ward [32]	Unknown	Unknown	US	27%
		Unknown	1200	US	14%
	Gulf Power Company [15]	Unknown	3000	US	22%
	SWALEC [15]	Unknown	100+	UK	25%
	Hydro One Network [11]	Unknown	153	Canada	7.6%
	Primen Inc. [12]	Unknown	61	US	TBD
	Country Energy AU [13]	Unknown	200	Australia	16%

[a] Conservation in terms of peak shifting.

TOU pricing policy being very effective when introduced with IHD, yields incremental conservation results. On the other hand, the prepay method also has its own benefits, but making a hybrid system with prepay method, TOU pricing policy, and introduced with IHD for energy monitoring is more likely to yield the highest amount of conservation among all categories.

Though the pilot projects have been very successful and most of them triggered a high positive response from consumers, they are mostly implemented in the United States, Canada, and Scandinavian countries. The question remains: Is the monitoring system feasible for the rest of the world? These projects were highly monitored and controlled over a specified and limited number of experimental groups, which makes its feasibility on a larger scale questionable.

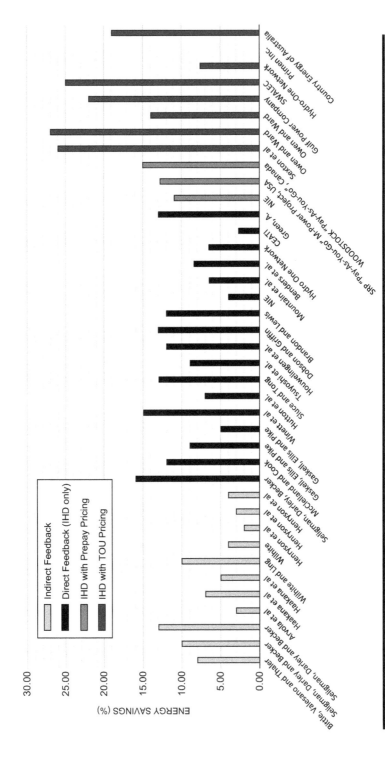

Figure 2.4   Comparison chart between different monitoring systems.

References

1. F. Al-Turjman and S. Alturjman, "Context-sensitive Access in Industrial Internet of Things (IIoT) Healthcare Applications", *IEEE Transactions on Industrial Informatics*, vol. 14, no. 6 (2018): 2736–2744.
2. "Honeywell Microelectronics and Precision Sensors", Neurophysics.ucsd.edu. [Online]. Available: https://neurophysics.ucsd.edu/Manuals/Honeywell/HMC%201 001%20and%20HMC%201002.pdf
3. Hydro One Networks, "The Impact of Real-Time Feedback on Residential Electricity Consumption: The Hydro One Pilot," Toronto, Ontario, 2006.
4. CEATI International, Inc., *"Results of Two-year Study Demonstrates Residential Electricity Monitors Help Homeowners Conserve Electricity in a Big Way."* Clearwater, Florida and St. John's, Newfoundland and Labrador, 2008.
5. A. Green, *"Potential of In-Home Displays in the PG&E Service Territory,"* Pacific gas and electric company, 2008.
6. T. Ueno, F. Sano, O. Saeki, and K. Tsuji, "Effectiveness of an Energy-Consumption Information System on Energy Savings in Residential Houses Based on Monitored Data", *Applied Energy*, vol. 83, no. 2 (2006): 166–183.
7. B. Pruitt, "SRP M-Power: A Better Way to Keep Customers in Power & Save Energy.", 2005.
8. F. Al-Turjman, "Energy–aware Data Delivery Framework for Safety-Oriented Mobile IoT," *IEEE Sensors Journal*, vol. 18, no. 1, pp. 470–478, 2007.
9. "SRP Wins NERO Energy-Efficiency Award, Washington, DC and Phoenix, Arizona," Metering.com, 2018. [Online]. Available: http://metering.com.
10. K. Quesnelle, *"Pay-As-You-Go-Power, Treating Electricity As a Commodity.",* Woodstock, Ontario: Woodstock H, 2004.
11. Hydro One Networks Inc., *"Time-of-use Pricing Pilot Project Results,"* Toronto, Ontario: Hydro One Networks, 2008.
12. Primen, Inc., "Final Report, Information Display Pilot, California Statewide Pricing Pilot," California, 2004.
13. CEAU, "The Country Energy Home Energy Efficiency Trial," *Sustainability Victoria*, New South Wales, Australia: State Government, 2006.
14. A. Faruqui, S. Sergici and A. Sharif, "The Impact of Informational Feedback on Energy Consumption—A Survey of the Experimental Evidence,", *Energy*, vol. 35, no. 4 (2010): 1598–1608.
15. S. Darby, "Making it Obvious: Designing Feedback into Energy Consumption," *Energy Efficiency in Household Appliances and Lighting.*Springer, Berlin: Heidelberg, (2001): 685–696.
16. R. Bittle, R. Valesano, and G. Thaler, "The Effects of Daily Feedback on Residential Electricity Usage as a Function of Usage Level and Type of Feedback Information," *Journal of Environmental Systems*, vol. 9, no. 3 (1979): 275–287.
17. C. Seligman, J. Darley, and L. Becker, "Behavioral Approaches to Residential Energy Conservation,", *Energy and Buildings*, vol. 1, no. 3 (1978): 325–337.
18. A. Arvola, A. Uutela and U. Anttila, "Billing feedback as a means of encouraging conservation of electricity in households: A field experiment in Helsinki", in *Energy and the Consumers*, pp. 75–85, 1994.

19. M. Haakana, L. Sillanpää, and M. Talsi, "The Effect of Feedback and Focused Advice on Household Energy Consumption," in *Proceedings, European Council for an Energy-Efficient Economy*, 1997.
20. H. Wilhite and R. Ling, "Measured Energy Savings From a More Informative Energy Bill," *Energy and Buildings*, vol. 22, no. 2 (1995): 145–155.
21. H. Wilhite, *"Experiences with the Implementation of An Informative Energy Bill in Norway,"* Ressurskonsult Report 750, 1997.
22. J. Henryson, T. Håkansson, and J. Pyrko, "Energy Efficiency in Buildings through Information—Swedish Perspective," *Energy Policy*, vol. 28, no. 3(2000): 169–180.
23. L. McClelland and S. Cook, "Energy Conservation Effects of Continuous In-Home Feedback in All-Electric Homes", *Journal of Environmental Systems*, vol. 9, no. 2, pp. 169–173, 1979.
24. G. Gaskell, P. Ellis, and R. Pike, *"The Energy Literate Consumer: The Effects of Consumption Feedback and Information on Beliefs, Knowledge and Behaviour."* London; London School of Economics and Political Science Department of Social Psychology, 1980.
25. R. Winett, J. Kagel, R. Battalio, and R. Winkler, "Effects of Monetary Rebates, Feedback, and Information on Residential Electricity Conservation," *Journal of Applied Psychology*, vol. 63, no. 1 (1978): 73–80.
26. R. Hutton, G. Mauser, P. Filiatrault, and O. Ahtola, "Effects of Cost-Related Feedback on Consumer Knowledge and Consumption Behavior: A Field Experimental Approach," *Journal of Consumer Research*, vol. 13, no. 3 (1986): 327.
27. A. Sluce and D. Tong, *"Energy Efficient Refurbishment of Victorian Terraced Housing: A Demonstration for Merseyside Improved Houses, Liverpool."* BRECSU Energy Studies, Final Report no. ED 227/209, 1987.
28. J. van Houwelingen and W. van Raaij, "The Effect of Goal-Setting and Daily Electronic Feedback on In-Home Energy Use," *Journal of Consumer Research*, vol. 16, no. 1 (1989): 98–105.
29. G. Brandon and A. Lewis, "Reducing Household Energy Consumption: A Qualitative and Quantitative Field Study," *Journal of Environmental Psychology*, vol. 19, no. 1 (1999): 75–85.
30. D. Mountain, *"The Impact of Real-Time Feedback on Residential Electricity Consumption: The Hydro One Pilot."* Ontario: Mountain Economic Consulting and Associates Inc., 2018.
31. R. Benders, R. Kok, H. Moll, G. Wiersma, and K. Noorman, "New Approaches for Household Energy Conservation—in Search of Personal Household Energy Budgets and Energy Reduction Options," *Energy Policy*, vol. 34, no. 18 (2006): 3612–3622.
32. G. Owen and J. Ward, *"Smart Meters: Commercial, Policy and Regulatory Drivers,"* London: Sustainability First, 2006.
33. G. Singh and F. Al-Turjman, "A Data Delivery Framework for Cognitive Information-Centric Sensor Networks in Smart Outdoor Monitoring", *Elsevier Computer Communications Journal*, vol. 74, no. 1 (2016): 38–51.
34. F. Al-Turjman, H. Hassanein, and M. Ibnkahla, "Towards prolonged lifetime for deployed WSNs in outdoor environment monitoring", *Elsevier Ad Hoc Networks Journal*, vol. 24, no. A (2015): 172–185.

35. F. Al-Turjman and A. Radwan, "Data Delivery in Wireless Multimedia Sensor Networks: Challenging & Defying in the IoT Era", *IEEE Wireless Communications Magazine*, vol. 24, no. 5 (2017): 126–131.

36. M. Z. Hasan, H. Al-Rizzo and F. Al-Turjman, "A Survey on Multipath Routing Protocols for QoS Assurances in Real-Time Multimedia Wireless Sensor Networks", *IEEE Communications Surveys and Tutorials*, vol. 19, no. 3 (2017): 1424–1456.

Chapter 3

Smart Homes in the Crowd of IoT-based Cities

Fadi Al-Turjman* and Chadi Altrjman†

Contents

* Antalya Bilim University, Antalya, Turkey
† University of Waterloo, ON, Canada

3.1 Introduction

A smart home is an application enabled by ubiquitous computing in which the home environment is monitored by ambient intelligence to provide context-aware services and facilitate remote home control [1]. Furthermore, it is considered to be a combination of several enabling technologies such as sensors, multimedia devices, communication protocols, and systems. From a different perspective, a smart home is merely a residence equipped with different Internet-connected devices that are used to remotely monitor and manage the appliances and systems installed in the home, such as lighting and heating, to mention just a few examples. Such a smart residence would be useful in managing the daily lives of the inhabitants. With the recent developments in the Information and Communication Technologies (ICT) and the reduction in the costs of low-powered electronics, a new technology has drawn the attention of the research community, namely the Internet of Things (IoT).

IoT is a revolutionizing technology, intending to connect the entire world by connecting physical smart devices used for sensing, processing, and actuating [2, 3]. By integrating Machine to Machine (M2M) communication technologies with the smart devices, these devices can connect and interact without any human intervention. As a result, IoT is believed to enable a fully conductive environment that can influence the life of society in different aspects such as the everyday activities of individuals and businesses, economy applications, healthcare applications, energy applications, the controlling of traffic and roads, and even political systems, to mention a few. Moreover, the "Things" are merely the devices and objects connected to a common interface with the ability to communicate with each other. By integrating the three core components of the IoT, namely the Internet, the things, and the connectivity, the value of IoT is to close the gap between both the physical and the digital worlds in the self-reinforcing and self-improving systems. The concept of smart homes is considered as an IoT-based application enabled by connecting the home appliances to the Internet. The home system's main goal is to provide security, monitoring, and control of all devices in homes over a cloud.

To achieve security, the system detects any threats in the home such as gas leaks, water leaks, and fires (it alarms the residents to prevent any losses of lives or properties). In addition, the system provides instant detection of any robberies happening. The controller manages all the devices installed in the home and it can remotely control these devices with the aid of smartphones. In addition, the system is compatible with all kinds of devices with the ability to manage their running time.

Amazon Web Server has built IoT specific services, such as AWS Greengrass, and AWS IoT. These services help people to collect and send data to the cloud, to load and analyze data, and to manage devices. AWS IoT is a managed cloud platform which allows the connected devices to easily and securely interact with each other and with cloud applications. AWS IoT is a managed service built for the purpose of connecting the devices to each other and to the cloud. Moreover, it can handle billions of devices and trillions of messages, with the ability to reliably and securely process and route these messages to the AWS endpoints and to other devices.

In this chapter, we will overview the IoT-based smart homes, the sensor types that can be deployed, the enabling standards, and the cloud architecture of a smart home. The rest of the chapter is constructed as follows. In Section 3.2, we will overview the different types of sensors used in smart homes. Section 3.3 discusses the communicating protocols and how they work. In Section 3.4, the common applications that can be deployed in smart homes are proposed. Section 3.5 discusses the cloud architecture for smart homes, and the chapter is concluded in Section 3.8. For better readability, the abbreviations used in this chapter are summarized in Table 3.1.

3.2 Smart Home Sensors

A sensor is an electronic component, module, or subsystem whose purpose is to detect events changing in its environment and send the information to other electronic devices, mostly a computer processor. A sensor detects and responds to some type of input from the physical environment. The output is generally a signal that is converted to a display that is readable by humans at the sensor location or transmitted electronically over a network for reading or further processing. Sensors in smart homes can be classified as follows in the next few sections.

3.2.1 Motion Sensors

This type is used in numerous applications including home security lights, automatic doors, and bathroom fixtures, that transmit some type of energy, such as microwaves, ultrasonic waves, or light beams, to mention a few. In addition, these sensors can detect when the energy flow is interrupted by something entering its path. These applications include:

Window and door control: This type of sensors monitors the doors and windows of the smart home and informs the owner about the people who entered or left their house. Moreover, these sensors can lower the energy consumption by switching the lights on and off when someone opens or closes the doors. The window-deployed sensors can be considered as the first line of defense from home break-ins. Some of these sensors detect any intruders trying to enter the house and alert the inhabitants. As these sensors are wirelessly connected to the Internet, the

Table 3.1   Acronyms Used in This Chapter

Acronym	Description
AWS	Amazon Web Server
PIR	Passive Infrared sensor
MW	Microwave
LED	Light Emitting Diode
UV	Ultraviolet
IoT	Internet of Things
AC	Air Conditioner
VRF	Variable Refrigerant Flow
VRV	Variable Refrigerant Volume
UPB	Universal Powerline Bus
BLE	Bluetooth Low Energy
ISM	The Industrial, Scientific, and Medical Radio Band
HVAC	Heating, Ventilation, and Air Conditioning
P2P	Peer-to-Peer
WSDL	The Web Services Description Language
SaaS	Software as a Service
PaaS	Platform as a Service
IaaS	Infrastructure as a Service
OSGi	Open Services Gateway Initiative
XML	Extensible Markup Language
SOAP	Simple Object Access Protocol
UDDI	Universal Description, Discovery, and Integration
HGW	Home Gateway
WLAN	Wireless Local Area Network
ULD	User-Friendly Location Discovery
OEM	Original Equipment Manufacturer

owner of the house can receive notifications on their smartphone and call for any help needed.

Glass Break Sensors: Once a window is broken the smart house sensor sends a signal back to the smart home panel.

Door contacts: Once armed, a signal is sent to the smart home control panel if the door is opened.

Video doorbell: This device protects the properties by working as a robbery hindering sensor. Also, it monitors anything behind the doors. This enables the house owner to know if anybody approaches their doors, whether they are at home or away from it.

Passive Infrared (PIR): This type of sensor detects infrared energy, so it can recognize any human near to it by sensing his body heat. These sensors are widely used in home security-based applications by detecting the heat and movement and then creating a protective grid. For instance, if a moving object blocks different zones, the infrared levels change and the sensors detect the movement and activate the protection grid. In [4], an application used for human localization in an indoor environment. This application uses the PIR sensors.

Microwave (MW): This kind of sensor sends out microwave pulses to measure the reflection of moving objects. MW sensors have an advantage over the infrared sensors in that they can cover more area. However, they are expensive and vulnerable to electrical interference. One of the common applications of microwave sensors is human identification in smart homes as proposed in [5].

Dual-Technology Motion Sensors: These sensors use multiple technologies such as PIR and MW sensors, to monitor a specific area. Both sensors must detect a signal to trigger the alarm or to activate a specific process. This can reduce the instances of false alarms.

Area Reflective Type: This type of sensor has a LED that emits infrared rays. The sensor detects an object if these radiated rays are reflected off it. Hence, the sensor measures the distance to the object and recognizes if the object is within the designated area

Ultrasonic sensors: They send out pulses of ultrasonic waves and measure the reflection off a moving object.

Vibration: These sensors detect vibration. There are two main sensor types in this category: The accelerometer and the piezoelectric device.

Photonic sensors: These sensors detect the presence of visible light, infrared transmission (IR), and/or ultraviolet (UV) energy.

3.2.2 Physical Sensors

These types of sensors are used in numerous applications listed as follows:

Temperature: Installed in beds and in some chairs to detect user location based on body temperature.

Tap sensor: This type of sensor uses infrared energy to sense the presence of an object. For instance, if the user puts their hands in front of the tap, the infrared sensor detects them and sends a signal to the solenoid valve to allow the flow of water. When the user moves their hands away from the tap, the sensor sends a signal to the valve to immediately terminate the water flow. This can help in saving water.

Humidity sensor: These sensors can be installed in bathrooms to detect any rapid increase in humidity levels which are defined by the users. The sensor sends these data to the Home Gateway (HGW).

Security: With an unlimited number of security sensors and settings, the residents can have full protection at home. While they are not home, they can start a pilot program to give the impression that they are at home. Moreover, these sensors can follow the intruders, and switch off/on the resident's lights depending on the motion. Also, while there is nobody at home, the security sensors can follow the kids at home to ensure their safety.

Flood: This kind of sensor protects your home and workplace from floods. With an early warning system of any water leakage within the house, the related valve can be shut down.

Gas leak: With the help of gas detectors, the sensor gets an alert for the slightest gas leaks. In addition, it will shut down the related valve. Thus, it will prevent any trouble that would occur due to such gas leakage.

Panic button: In emergency cases, this sensor would text or email your relatives or security institutions to send the help needed.

Shake sensor: In case of earthquakes an early warning system is activated. The system will carry out your predetermined actions.

Irrigation: These sensors can check your irrigation valves in your garden, create your own irrigation program, or determine automatic irrigation scenarios according to the weather.

Curtain control: These sensors can remotely control your curtains or canopy. You can also protect your home from extreme weather conditions using preprogrammed scenarios.

Multimedia control: Control your home theater from a distance with your smartphone.

Ventilation: Split AC, VRF, VRV, and any kind of ventilation system can be at your disposal, and you can program and control it.

TV control: Change channels or increase or decrease the volume using smartphones.

Smart plug: Control your plugged devices remotely. Measure energy consumption or automatically shut down devices even when you are not at home.

3.2.3 Chemical Sensors

These types of sensors are used in numerous applications listed as follows:

Fire/CO detection: Fire is considered as the main reasons for property damage. For years, the humble fire detector has been used for detecting any smoke in homes. Nevertheless, there are many pollutants threatening the air quality inside the smart homes, resulting in harm to the residents. One of these dangerous pollutants is carbon monoxide (CO) which can be sensed by a CO detector. This device can be employed in emergency monitoring service to sense this odorless gas. The new versions of these sensors have many features, one of them is that they can't only monitor pollutants such as dust, soot, pollen, and particulates, but also, they monitor the overall air quality such as the humidity and the air stakes. Moreover, the customer can receive attractive discounts from insurance companies when they use these sensors.

3.2.4 Leak/Moisture Detection

One of the main reasons for damage to the home is water flooding. For instance, if a water pipe is broken in the house and water flows from it continuously for many hours, this will destroy the devices and appliances in the home. A common solution for this is to deploy a moisture detection sensor to alarm the house owner if there is a water leak in it. These sensors can be placed near to water heaters, dishwashers, refrigerators, sinks, and sump pumps. Further, if the sensor detects any unwanted moisture, the smart system can send a notification to the house owner about the problem, to find a solution for it quickly.

3.2.5 Remote Sensors

Smart garage door: The Wi-Fi connected smart garage door is a device that helps the house owner in cases where they forget to close the garage door. It enables the user to remotely control the garage door via a smartphone.

Smart thermostat: This device allows the user to remotely control the heating and cooling system within the house. Furthermore, it can control the humidity and adjust the temperature of the house based on the user's behavior inside the house. Moreover, these sensors can activate the energy-saving mode when there are no users in the house. Applying cognitive technology to these sensors is paving the way for a smart home to be able to learn more about the residents and their temperature preferences.

3.2.6 Biosensors

Alert Z-Wave Smoke Detector and CO Alarm: When it detects smoke or CO, it sounds an alarm and sends an alert to your phone. It can also contact any other number of people, including the fire department, if you're away. The First Alert detector, namely Albert Z-Wave Smoke Detector, is a great option that works with your SmartThings (or other Z-wave) hub.

Smoke: These sensors are used for detecting smoke. They can be installed in the ceiling of each partition. If a user is smoking, their location is identified based on a signal sent by this sensor.

3.3 Protocols for Smart Home Devices

Protocols are how signals are sent from one device to another in order to trigger an action, such as turning the lights on and off. There are a wide variety of technology platforms, or protocols, on which a smart home can be built. Each one is, essentially, its own language. Each language is related to various connected devices. In addition, it instructs them to perform a function. Choosing a smart home protocol can be a tricky business. It is better to choose a protocol that can support a large number of devices, and offer the best possible device interoperability, i.e., the ability for devices to talk to each other. There are many factors to consider when choosing a protocol, such as power consumption, bandwidth and, of course, cost [6].

3.3.1 How Home Automation Protocols Work

Home automation protocols are considered to be the communication frames used by the smart home devices to interact with each other, due to the vital role of communication to achieve the required home automation. It involves automatic controlling of all electrical or electronic devices in homes or even remotely through wireless communication. Centralized control of lighting system, air conditioning and heating, audio/video systems, security systems, kitchen appliances, and all other equipment used in home systems are enabled by this kind of protocols. Moreover, implementation of the home automation depends on the type of controls, such as wired or wireless.

3.3.2 Differences Between Home Automation Protocols

There are mainly three types of home automation systems. Powerline Based Home Automation: This automation is inexpensive and doesn't require additional cables to transfer the information. Instead, it uses existing power lines to transfer the data. However, this system involves a large complexity and necessitates additional converter circuits and devices. Wired protocols such as UPB and X10, use the wiring existing in homes for communication. Although these protocols are reliable, they are slow and can face several difficulties to encrypt. On the other hand, wireless protocols usually tend to be faster and more congenial with many other devices. Wireless home automation protocols such as Z-Wave, ZigBee, Wi-Fi, Thread, and Bluetooth, are easier to be secured because they can communicate without using any power lines. Also, there are some protocols that utilize both wired and wireless communication such as C-Bus and Insteon. As a result, they avail from the features provided by both wired and wireless technologies. Moreover, a protocol's compatibility with smart devices is a measure of how many smart devices that the protocol can be easily employed with [7].

Z-Wave – This is seen as a wireless protocol working in the radio frequency band and used by many home automation devices to communicate with each other.

Z-Wave runs on the 908.42 MHz frequency band. This frequency band is much lower than the one used by most household wireless products (2.4 GHz). As a result, it is not affected by their interference and there are no traffic jams in this band. One significant advantage of Z-Wave is the interoperability of the devices connected to it which means that the devices communicate to all other Z-Wave devices, regardless of type, version, or brand. Also, the interoperability is backward- and forward-compatible in the Z-Wave ecosystem.

How It Works - In order to set up and manage the home automation network, Z-Wave uses one central hub. In addition, the user can easily add any smart home devices and manage them using the Z-Wave protocol once the network is installed [8].

Z-Wave Compatibility - With more than 1,700 certified Z-Wave devices available in the world, it is easy for homeowners to select from plenty of options for automating their homes. In addition, Z-Wave devices can be easily set up and use and consume less energy.

Benefits of Z-Wave - As mentioned before, the Z-wave uses a frequency band which is much lower than the frequencies used for most of the other wireless devices. As a result, there is a lower chance of interfering with other frequencies which are crowded with many competing devices. So, the Z-wave has a more efficient communication with home automation devices. Also, regardless of the type or the version of the devices, they can freely interact with each other. This works also for the continuously evolving product. So, every time these products get updated, the users won't start with a new hub. This has a special importance nowadays where everything is updating quickly.

ZigBee – This is a popular wireless home automation protocol which is the same as Z-Wave protocols in many ways. ZigBee was originally developed to be commercially used. However, it is now utilized as a standard protocol for many automation applications in both residential and commercial environments.

How It Works - Like the Z-wave protocol, the ZigBee protocol uses radio frequency to communicate. Furthermore, The ZigBee protocol is a low-cost, low-powered technology, which means that the battery-operated devices in a ZigBee will enjoy a long life. Moreover, it uses a mesh network which enables a larger range and faster communication. Also, One ZigBee hub can be used by multiple devices which makes it suitable for automating home appliances.

ZigBee Compatibility - Recently, the ZigBee technology is considered to be an open environment for developers to design many applications that can work with it. As a result, currently, there are more than 1200 products certified and compatible with a ZigBee hub. Early on, interoperability of the connected devices, which are made by several manufacturers, was a big challenge facing ZigBee. However, the recently released versions tend to have a better operability, regardless of their manufacturer or version.

Benefits of ZigBee - ZigBee has a low power usage, so the user can use a ZigBee device without changing the batteries for several years. Also, enabling ZigBee at home automation offers a green power feature which makes it unnecessary to use any batteries. For security issues, using the same level of encryption that most of the financial institutions use, it is seen as one of the best protocols regarding high-security levels. This high security will cover the ZigBee network, devices, and the information conveyed by it. Furthermore, ZigBee home automation is highly customizable and can be perfectly used by techies.

Insteon – The Insteon protocol is a hybrid technology, which uses both wireless and wired communication technologies. This makes it unique in the field of home automation.

How It Works - Insteon home automation runs on a patented dual-mesh network that uses both wired and wireless communication to overcome general difficulties facing each type while operating solely. An Insteon hub connects with Insteon-compatible devices. This creates the ability to control the user's smart home via smartphones.

Insteon Compatibility - With more than 200 Insteon-compatible home automation devices available commercially, Insteon tends to rely more on them, unlike many other protocols. This creates a restricted compatibility with smart devices made by other manufacturers. On the other hand, Insteon--compatible devices have a feature to enable forward and backward compatibility, meaning that they can work well with both new and old versions of devices.

Benefits of Insteon - Regardless of what the user's level of technical prowess is, they can run Insteon home automation with no running problems. The user need only know how a smartphone is used and the Insteon protocol is easy to manage. Moreover, as the user turns on any Insteon-compatible devices, they can be automatically added to the network. This makes them speedy devices to setup and the user faces no trouble with connecting them. In addition, the Insteon protocol has a notable feature regarding the network size, the network is not limited, and it can be as large as the user desires. This is because the network is a dual-mesh and one Insteon hub is able to work with hundreds of devices across a broad range without facing any problems.

Bluetooth – This is considered to be the core communication frame for hundreds of smart products nowadays. Having a higher data bandwidth than ZigBee and Z-Wave, made it suitable for home automation applications. However, its bandwidth is lower than the Wi-Fi, it consumes far less power than Wi-Fi. Bluetooth Smart, or Bluetooth Low Energy (BLE), is an energy efficient version of Bluetooth wireless technology often seen in smartphones, and ideal for use when the smartphone is connected to a headset. Because it is an energy efficient protocol and compatible with the existing smart devices, it is easy for developers and OEMs to create solutions that can immediately be added to the existing systems. Moreover, although it uses different channels, BLE operates in the already crowded 2.400 GHz–2.4835 GHz ISM band and the required data compression diminishes the audio quality. Also, BLE can be used within the home due to its low energy usage which can extend the battery life of the devices [9].

How It Works - Bluetooth is a wireless home automation protocol that uses radio frequencies for communication. Containing a computer chip with a Bluetooth radio and a software makes it simple for various devices to interact with each other. Moreover, the users can use one primary Bluetooth hub for controlling all the devices connected to their home automation network.

Bluetooth Compatibility - With the ability to connect Bluetooth-enabled devices with each other, hundreds of Bluetooth-compatible products commercially exist now. Nevertheless, other non-Bluetooth devices can't be added to such a Bluetooth hub. Moreover, Bluetooth connections have a limited range connection. As a result, many congenial devices can fail to work in such a network because of the limited range of its communication.

Benefits of Bluetooth - Bluetooth is one of the fastest-growing sectors of home automation and it is included in many devices. This is due to the fact that its home automation products are desired in many applications and consume little running power, resulting in reducing the carbon footprint.

UPB – This stands for universal power line bus. It is a home automation protocol that uses wired communication technology. UPB was released in 1999 and is considered to be one of the most technically advanced protocols.

How It Works - Based on the X10 standard, the UPB home automation protocol uses the power lines for enabling communication between different devices. UPB devices can connect to the network using two enabling devices, a central home controller, which is manually set up by the user, and links for each device connected to the network. Also, UPB protocol doesn't require too much technical savvy for the users to set up and run.

UPB Compatibility - Compared to the other protocols, UPB has fewer home automation products available at present. One reason for this is that UPB is difficult to combine with any wireless protocols. On contrast, there are around 150 commercial UPB- compatible products existing in the market.

Benefits of UPB - UPB is one of the most reliable protocols. This is because it is hard-wired into the power lines, so very limited interference is experienced compared to wireless home automation technologies.

Thread – This is a new wireless protocol for smart household devices. The Thread Group was formed in July 2014 by seven founding members, including Google's Nest Labs and Samsung Electronics. More than 250 devices can be connected to a Thread network. Also, because most of the devices meant to be connected to the network are battery-operated, it's very frugal on power. Using the same frequency and radio chips as ZigBee, Thread is intended to provide a reliable low-powered, self-healing, and secure network that makes it simple for people to connect more than 250 devices in the home to each other. Moreover, they can be connected to the Cloud for ubiquitous access. Nest Learning Thermostat and Nest Protect are already using a version of Thread, and more products are supposed to enter the market soon.

3.4 Applications

Many applications that can be employed in smart homes are proposed in [10]. The most common applications of home automation are lighting control, HVAC, outdoor lawn irrigation, kitchen appliances, and security systems. An occupancy sensor adjusts the temperature and turns off lights when a room is not in use. Window contacts setback HVAC when windows or balcony doors are left open. Wall switches are used to control lighting and shading. A heat valve is used for self-powered and energy efficient room temperature control. A room temperature sensor is used for minimal energy consumption and maximum comfort. A plug-in receiver controls and monitors consumer appliances. Motion activated sensors, as the name implies, these home security sensors are used to detect an intruder's presence. They are used for smart home protection. Once a sensor is tripped by a thief entering your home, a signal is sent to the smart home panel. In addition to setting off the internal siren, the smart home panel will also send you a text message or email, informing you of events as they unfold. Door contacts, once armed, if the door is opened a signal is sent to the smart home control panel.

3.5 Cloud Architecture for Smart Homes

Cloud Computing is a recently introduced technology trend with the aim of delivering computing facilities as Internet services. There are many thriving commercial Cloud services examples built by existing companies such as SaaS, PaaS, and IaaS, to merely mention a few. However, these services are all computer-based and mainly designed for the services provided by web browsers. Currently, we cannot find a cloud architecture that can provide the users with special services to the digital devices installed in smart homes. In this study, we propose an additional cloud model, namely the smart home Cloud, which relies on the current cloud architecture with a modification of the service layer. This would result in providing efficient and stable services for the owners of smart homes. On the other hand, we recommend a Web service and Peer-to-Peer (P2P) technologies (see Figures 3.1 and 3.2) to connect to the Cloud. This way, the cloud server would be able to provide a better service for the higher-quality audio/video signals by reducing the bandwidth pressure while transmitting them. In addition, the smart HGW is responsible for both describing their services in WSDL and registering them in the cloud service directory. This way, the other homes can search for the service and benefit from it. Consequently, we can consider smart homes as service consumers and suppliers at the same time. P2P and Web service are reasonable technologies that can be introduced to the cloud for combining both the cloud and smart homes. By using these two technologies, the cloud successfully provides a more specialized functionality to the smart homes. Recently, there has been too much research focusing on the HGW. One of the initiatives that this resulted in is the Open Services Gateway

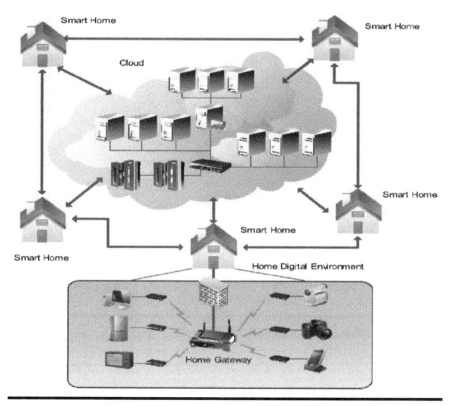

Figure 3.1 P2P network diagram.

Initiative (OSGi). This initiative intends to create a platform that enables deploying services over both the wide-range network and local network or device. In the OSGI architecture-based smart home, the residential service gateway has two main functions; connecting the appliances inside the house and linking the external network to the outside of the house. This way, people can experience a better home life without the need to overpower the smart devices with complex technologies and spontaneous user interfaces [11].

3.5.1 Peer-to-Peer Networks

In this part, we consider the smart home services as classified into three main categories, namely, home entertainment, video communication, and video conferencing, which are all based on audio/video streaming. The traditional client-server cloud model is constrained by the problem of bandwidth load pressure while transferring heavy high-quality audio/video signals. Solving such a problem required us to propose a mechanism where a Cloud server and the smart home nodes can set a P2P network. This will reduce the bandwidth loads on the cloud. For this model, each user will be

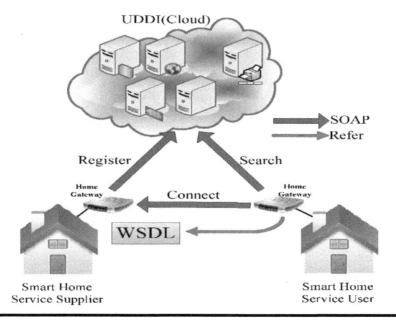

Figure 3.2 Web service architecture diagram.

responsible for registering some information to the cloud such as his name and IP. Moreover, a specific software for the P2P communication should be installed in the gateways of each home. As a result, the home nodes can share the bandwidth loads and the overall processing power. For instance, a user can set a public cloud to discern the real-time HD video broadcast for the other users who share the same cloud with him. Also, the other users can benefit from such a service by watching this video. So, the participant who is broadcasting the video stream is a customer downloading the video and a seller, who is uploading it to the others, at the same time.

3.5.2 Web Service

In the traditional client-server model, the homes are considered as consumers of services only. In contrast, in our model, the smart homes (Peers) are consuming services and sharing them with the others. So, we can consider them as both consumers and suppliers of a service at the same time. To understand this feature, a Web service technology is introduced [11]. It uses Extensible Markup Language (XML) messages and follows the Simple Object Access Protocol (SOAP) standard. The HGW is responsible for connecting and manages the various devices and networks installed within the house such as the home automation network and the PC network. As a result, this can create an integrated digital environment. Moreover, this model enables the users to use the HGW to share some features provided by their appliances with the neighboring houses. This can be achieved by describing the required services in a Web Services Description Language (WSDL) and distributing them into the cloud. Once the services are shared in the cloud, the other users can search for and use them. Hence, the cloud is considered as a UDDI server in the system, in which the service providers can publish their services and the customers can discover them. Also, it allows them to define how the services and the applications interact with each other on the Internet.

In Figure 3.2, the proposed Web service in the smart home architecture is shown. For instance, if a user has a printer in smart home A and they wants to share one of its services with the other houses, the house service gateway construes the service related to the online printing in WSDL and registers it to the Service Directory in the Cloud. Other smart houses can benefit from this service after paying to the owner of Home A. So, smart home A can make a profit from publishing his service on the cloud.

3.6 Architecture of Smart-Home-Based Cloud

The cloud architecture can be divided into three parts, namely, the platform layer, the infrastructure layer, and the service layer as shown in Figure 3.3. The platform layer with infrastructure layer creates the base for delivering traditional IaaS and PaaS. On the other hand, the service layer is responsible for directly interacting with the smart home and focuses on the application services (SaaS).

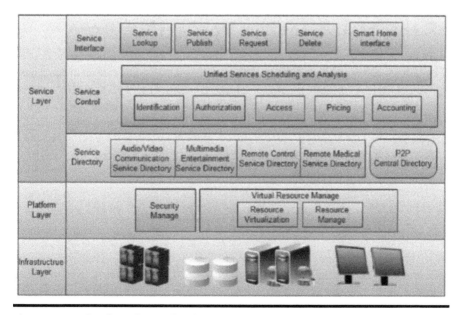

Figure 3.3   The three-layer cloud hierarchy for smart homes.

3.6.1 Service Layer

This part includes three main parts, namely, the service interface, the service directory, and the service control. The service interface directly interacts with the smart home users. Furthermore, the users can easily publish their service description using the service interface. The service control part, which is considered to be the brain of the service layer, is responsible for analyzing, processing, and responding to the user's requests.

3.6.2 Platform Layer

This layer is considered as the essence of the cloud model presented in the previous section. It includes two main modules, specifically, the security manage module and the resources management module. Also, the Platform Layer with the help of the Infrastructure Layer can create the base for delivering PaaS to smart homes.

3.6.3 Infrastructure Layer

The Infrastructure Layer contains a huge amount of physical resources responsible for delivering the cloud services. These resources are handled by a higher-level virtualization component, which is controlled by the cloud to provide IaaS Services for smart homes.

3.7 User-Friendly Location Discovery (ULD)

To improve the user's privacy and comfort, the ULD system is a must and is designed to be more user-friendly than conventional localization approaches. Figure 3.4 shows a high-level architecture for supporting context-aware services in future smart homes. Also, it clearly presents the role of our ULD system in the architecture. Several types of sensors monitor the home environment to detect the user(s). Each sensor that detects a user sends a witness signal (as context information) through a sensor network via heterogeneous interfaces (e.g., ZigBee, Bluetooth, WLAN) to a HGW.

3.8 Conclusion

Smart Home applications provide its homeowners with comfort, security, energy efficiency (low operating costs), and convenience. The smart home system offers solutions to problems such as: fire, flood, gas leak, theft, child/elderly care, smart

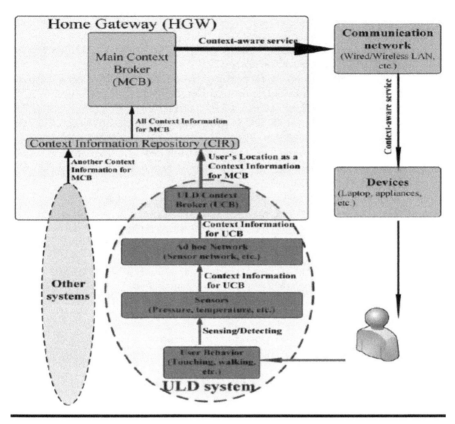

Figure 3.4 **A high-level architecture for supporting context-aware services in future smart homes [12].**

door, temperature and humidity control, security management, air conditioning and irrigation control, curtain-shutter control, light controller etc. Any devices can work in harmony with each other to improve quality of life. In addition to more security, the smart home system allows you to administer electrical devices such as TV etc. remotely. Moreover, it provides you with reports to prevent further energy consumption. You can control all these high security and comfort services from wherever you have Internet access with a smartphone, tablet, or computer. For the user, these services will be time efficient and more secure, because humans cannot remember everything. Therefore, the automated system can control all housework in an instant of time. For example, in terms of security management, smart homes are very useful. Thanks to smart home applications, users can regulate security management. Therefore, it will be by remote control. It will be cost-efficient and safer.

References

1. D. J. Cook, A. S. Crandall, B. L. Thomas, and N. C. Krishnan, "CASAS: A Smart Home in a Box," *Computer*, vol. 46, no. 7 (2013): 62–69.
2. A. Al-Fuqaha, M. Guizani, M. Mohammadi, M. Aledhari, and M. Ayyash, " Internet of Things: A survey on Enabling Technologies, Protocols, and Applications, " *IEEE Communications Surveys & Tutorials* vol. 17, no. 4 (2015): 2347–2376.
3. R. Want, B. N. Schilit, and S. Jenson, "Enabling the Internet of Things". *Computer*, vol. 48, no. 1 (2015): 28–35.
4. D. Yang, B. Xu, K. Rao, and W. Sheng, "Passive infrared (PIR)-Based Indoor Position Tracking for Smart Homes Using Accessibility Maps and A-Star Algorithm,". *Sensors*, vol. 18, no. 2 (2018): 332.
5. D. Sasakawa, N. Honma, T. Nakayama, and S. Iizuka, "Human Identification Using MIMO Array," *IEEE Sensors Journal*, vol. 18, no. 8, pp. 3183–3189, 2018.
6. F. Al-Turjman, "QoS -aware Data Delivery Framework for Safety-inspired Multimedia in Integrated Vehicular-IoT", *Elsevier Computer Communications Journal*, vol. 121, pp. 33–43, 2018.
7. Brian Ray, "Applications of Home Automation". Published March 17 2015. https://www.linklabs.com/blog/applications-of-home-automation
8. M. B. Yassein, W. Mardini, and A. Khalil, "Smart Homes Automation Using Z-Wave Protocol," *IEEE International Conference on Engineering & MIS (ICEMIS'16)*, IEEE, (2016): 1–6.
9. M. Siekkinen, M. Hiienkari, J. K. Nurminen, and J. Nieminen, "How Low Energy is Bluetooth Low Energy? Comparative Measurements with Zigbee/802.15,". *IEEE Wireless Communications and Networking Conference Workshops (WCNCW'12)*, IEEE, (2012): 232–237.
10. P. Gaikwad, J. P. Gabhane, and S. S. Golait, "A Survey Based on Smart Homes System Using Internet-of-Things," *IEEE International Conference on Computation of Power, Energy Information and Commuincation (ICCPEIC'15)*, IEEE, (2015): 0330–0335.
11. E. Newcomer, *Understanding Web Services: XML Wsdl Soap and UDDI*, Addison-Wesley Professional, 2002.
12. Ahvar, E., Lee, G. M., Han, S. N., Crespi, N., andI., Khan, I., Sensor Network-Based and User-Friendly User Location Discovery for Future Smart Homes, *Sensors*, vol. 16 no. 7 (2016): 969.

Chapter 4

Smart-Grid and Solar Energy Harvesting in IoT-based Cities

Fadi Al-Turjman* and Abdulsalam Abdulkadir†

Contents

* Antalya Bilim University, Antalya, Turkey
† Middle East Technical University, Ankara, Turkey

45

4.1 Introduction

In the twenty-first century, electricity utilization has changed significantly due to the huge increase in demand. Several uses of the main resources of electricity production surfaced. Hence the grid of the twentieth century is inadequate now. The constantly increasing demand for better and effective provision of power resulted in the development of a more robust, effective, and a two-way grid system called "smart grid" [1, 2]. Smart grid can be defined as the power generation network that intelligently integrates the utility generators and the end users to efficiently transmit and distribute electricity. It also allows electricity distribution to be sufficient in capacity with good area coverage and to be safe, economic, reliable, efficient, and sustainable. [1–3].

Also, smart grid is an interconnected network for delivering electricity from producers to end users. It consists of generating stations, high voltage transmission lines, and distribution lines. Issues resulting from connectivity are solved in practical scenarios using generic approaches [4], against machinery or human mistakes.

Solar power is radiant energy received from the Sun that is processed using a range of continuously changing technologies; this includes photovoltaic (PV), solar thermal energy, solar heating, solar architecture, molten salt power plants, and artificial photosynthesis [5]. Solar Energy is the changing of radiant energy collected from sun into electric power, either using PV directly, indirectly using concentrated solar panels (CSP) or using a combination of both. Concentrated solar systems use lenses or mirrors and tracking systems to focus a large area of sunlight into a small beam.

Internet of Things (IoT) is a term mostly given to a set of devices networked together to sense and collect data from everywhere. The origin of the IoT started in 1999 when Kevin Ashton first coined the name in the context of supply chain management. Initially the definition was usually used covering lots of ranges that involve applications in healthcare, utilities, transport, etc. but later changes as technology evolved, the focus being to make computers sense information without the help of humans. When home appliances are connected to a network, they can interact together in cooperation to give the ideal service as a whole and not as an individual entity together working as a single device. This concept referred to as the IoT became very acceptable, useful, and gained applications in many areas of human endeavor which include:

- FOOD as smarter food
- TRAFIC as smarter transportation
- ROAD as smarter roads
- HEALTH as smarter healthcare
- BLOOD BANK as smarter blood banks
- GRID as smarter grid etc

In the following section, we summarize our main contributions and abbreviations used. This chapter overviews potential methods in improving the energy generation process under varying conditions of scarcity and uncertainty in the IoT era. It considers optimization aspects of the self-generated power using roof-top solar panels and local energy storage facilities. We discuss the grid evolution, the different types of this, advantages, and what it requires to integrate with the generated solar energy using smart meters, tariff applications, storage devices, and optimization of storage infrastructure. Table 4.1 shows the list of abbreviations used in this chapter.

4.2 Literature Review

The first power grid was built in 1895 with the main aim of supplying electricity. Presently there are 9,200 power grids all over the world producing and supplying consumers with about 1 million MW of electricity on a daily basis [1]. The grid started from just a conventional one and developed to a Micro grid and the smart grid. The conventional grid was developed for conventional power generation from fossil fuel generating plants, nuclear plants, hydro power plants with significantly high power loss due to long transmission lines while the others were as a result of advancement in electricity production from solar and wind sources with the optimum aim of effective power transmission. The Micro grid is quite small which focuses more on local electrification and can stand outside the conventional grid or integrate with it. Its transmission of electricity is across short distances, therefore

Table 4.1   Abbreviations Used in the Chapter

Abbreviation	Description
TWh	Terawatt hour
GHG	Green House Gas
PV	Photovoltaic
CSP	Concentrated Solar Panels
IoT	Internet of Things
KW	Kilowatt
MW	Megawatt
FLISR	Fault Location Isolation and Service Restoration
FACTS	Flexible Alternating Current Transmission System
RES	Renewable Energy Sources
AMI	Advance Metering Infrastructure
AMR	Automatic Meter Reading
WAN	Wide Area Network
HAN	Home Area Network
3G	Third Generations
GSM	Global System for Mobile
NIST	National Institute for Standards and Technology
ToU	Time of Use
CAES	Compressed Air Energy Storage
NAS	Sodium Sulfur
EVs	Electric Vehicles
HEVs	Hybrid Electric Vehicles
FHEV	Full Hybrid Electric Vehicle
MHEV	Mild Hybrid Electric Vehicle
BEV	Battery Electric Vehicle
FCEV	Fuel Cell Electric Vehicle
PHEV	Plug-in Hybrid Electric Vehicle
MHV	Micro Hybrid Vehicle
V2G	Vehicle-to-Grid

loss in energy encountered during transmission in the conventional grid is much reduced with capacity a range of 3 kW–10 MW [6].

In Nigeria electricity generation started as early as 1896, with the installation of its first power plant of just 60 kW generating capacity that was more than the energy needed then, in Lagos in 1898 [7, 8]. Then with a series of expansions followed by neglect of infrastructural development, reported vandalizations, and of course corruption until today, the energy need for the continuously increasing population was never satisfied even after huge spending by different administrations. Presently, officials of the Administration of President Muhammadu Buhari reported electricity availability of 7000 MW, which is quite poor for approximately 186 million citizens (as of 2016) but available reports indicated just above half of that value (3,941 MW) is available from the 23 power plants connected to the National Grid as of May, 2015.

4.2.1 Smart Grid

Smart grid literally means upgrading the traditional grid from a one-way source to a two-way source of electricity generation, which is well digitalized, modernizing with well detailed automation for the twenty-first century electricity transmission and distribution by simply adding censors and software or programs to the conventional grid that enriches providers of electricity and end users with lots of information to work with in effective transmission and distribution of the electricity and certainly effective, low-cost utilization of the electricity by providing more knowledge of when electricity is cheap to use and when it is expensive to use.

Smart grid systems can be described as a combination of distributed and centralized generators which are utilized to control the low- and high-level voltage distribution through automation systems of the different users, bringing better secured and reliable energy to these users for their respective uses [1]. Smart grid is referred by [9], as a system that recently evolved using available Heterogeneous Networks (HetNets) to control our huge electricity needs in a smart, economical, and sustainable way. This creates competition in the energy market allowing different sources, such as solar and wind energy sources to include better monitoring of transmission facilities over regions [6] and most importantly it allows the possibility of end users to act as a service provider [10].

4.3 Advancements in the Smart Grid

The integration of the traditional electric grid with communications and information systems to better monitor and control the flow and consumption of energy; save energy, reduced cost and increase reliability of electricity generation and transmission that makes it a modern electric grid which makes it possible to connect electricity generated from non-conventional sources.

In a smart grid system sensors are deployed linearly and programs such as Fault Location Isolation and Service Restoration (FLISR), a distribution automation tool that helps in the optimization plan for switching action as well as restoring supply, are used, as well as a Flexible Alternating Current Transmission System (FACTS), that helps in a controllability enhancement including an increase in power capability that will send a notification if a grid shuts down and the service provider will be notified with all details regarding such [11]. It was reported that the integration of Renewable Energy Sources (RES) helped significantly in reducing system losses and increased the reliability, efficiency, and security of smart grid as well as making its infrastructure fully advanced in sensing and communication computing abilities. Hence it can be said that smart grid reduces Green House Gas (GHG) emission by allowing the integration of renewable sources to the grid and thus the dependence on fossil fuel decreased. The smart grid uses the synchrophasor technology to prevent constraint, systems collapse, natural incidents, and equipment failures which are the general conditions that cause energy disturbance. It can thus be said that this system can understand its situation and know in advance the challenges, thereby resolving crises before they even occur [1, 2].

The enhanced benefit of the smart grid is in smart metering that provides a two-way advantage to both providers and end users. The smart meter provides the service providers with detailed hourly utilization of energy by end users and uses the collected data in billing the homeowner as well as to balance the consumption of power by reducing the price for the homeowner at certain hours of the day to control peak energy consumption. Mainly a smart meter consists of the following components; analog and digital ports, serial communication, volatile memory, non-volatile memory, real-time clock, and microcontrollers.

4.3.1 Benefits of the Smart Grid to Energy Sector

The benefits of the twenty-first century grid referred to as a smart grid are numerous, most importantly being in the Advance Metering Infrastructure (AMI). Some added incentives provided by upgrading to smart grid are

- Provides customers the opportunity to have an effective financial plan on electricity usage
- Reliable electricity provision
- The provision of efficient, robust, and high-quality electricity
- Generation of more efficient renewable power
- Enhances energy sources mix
- Creates interface for smart devices in home automation
- Reduces carbon footprint
- Creates enabling environment for electric vehicles (EVs)
- Creates lots of job opportunities
- Anticipates faults or problems
- Self-healing response ability
- Provides alternative for storage
- Security of infrastructure

4.4 Metering Technologies

The meters used initially in electricity transmission, distribution, and supply were actually to allow a means of knowing the amount of consumed electricity for billing purposes only. These meters are referred to as Traditional meters and the technologies used by such meters are called Electromechanical Technology. The need to provide more efficient and reliable electricity resulted in the development of smart meters, such as Automatic Meter Reading (AMR) meters and later Advanced Metering Infrastructure (AMI) meters. The different types of meters from Traditional, AMR, to AMI meters are shown in Figure 4.1. The AMR meters use the electromechanical plus electronics technology that allows providers to read electricity consumption remotely using fixed network abilities such as handheld, walk by, or drive-by devices. The AMR still did not provide the much needed two-way advantage that allows consumers active participation, and this brought the advent of the AMI using the newest technology in metering referred to as Hybrid Technology.

4.4.1 Pros of Smart Meters versus the Conventional Meter

- The conventional meter can only provide usage data to the service provider, in most cases, collected manually for monthly billing.
- On the other hand, smart meter provides what is referred to as two-way information that helps the utility provider and the end user.
- It informs both parties of any power outage and possible theft.
- It also improves power quality and provides more efficient, secure power delivery, enhanced electricity reliability, allows prepayment options, and features more accurate billing,
- In short, it brings the end of estimated bills and over paying/underpaying.

Figure 4.1   Different types of meters.

4.5 Communications Used in Smart Grid

The smart meter is connected in two ways using Wide Area Network (WAN) and Home Area Network (HAN). WAN is used to connect the smart meter, the supplier and the utility server, while in connecting the smart meter with home appliances HAN is used. The different technologies used by WAN for communications are fiber optics or 3G/GSM while technologies used as a HAN are Bluetooth, ZigBee, and wireless Ethernet or wired Ethernet.

4.5.1 ZigBee

In wireless communication technologies, National Institute for Standards and Technology (NIST), confirmed ZigBee Smart Energy Profile as the most applicable communication infrastructure for smart grid network due to its low cost, comparatively low power consumption, complexity, and data rate transfer. ZigBee is used in smart grid for automatics meter reading, energy monitoring, and home automation. It has a bandwidth of 2.4 GHz plus 16 channels, every channel uses 5 MHz bandwidth and a maximum output power of 0 dBM including a transmission range of between 1 m–100 m with a data rate of 250 kb/s [1].

4.5.2 Wireless Mesh

A Wireless Mesh network is a combination of nodes that are joined in groups and work as a self-reliant router. The self-healing property of these nodes are helpful for a communication signal to find a route through active nodes. Infrastructures of mesh network are decentralized because each node sends information to the next node. Wireless mesh is used in small business operation and remote areas for affordable connections [1, 11, 12].

4.5.3 GSM

GSM is Global System for Mobile communication that is used to transfer data and voice services in communication technology. GSM is a cellular technology that connects mobile phone with the cellular network [1].

4.5.4 Cellular Network

Cellular networks can be another excellent option for communication between far nodes for utility purpose. Cellular networks are used to build a dedicated path for communication infrastructure to enable smart meter deployment over a WAN [1].

4.6 Billing Methods

There are a few main types of billing in the smart grid system. They are net metering, feed in tariff and time of use (ToU).

4.6.1 Net Metering

This method is used to measure the amount of extra energy a homeowner was able to produce to the smart grid. This extra saved amount is rewarded back to the user at the same rate during the night period or when the sunlight is insufficient to produce enough energy.

4.6.2 Feed in Tariff

In this situation a homeowner pays a rate which is usually calculated using the ToU. Should he need more energy back it will be sold to him at a similar rate.

4.6.3 Time of Use

This is a billing plan in which the electricity consumption is calculated based on the real time in which such energy is utilized. The pricing of electricity usage is significantly lower when used at a time that the demand for power supply is low and higher when used at a time or period of high power demand. When a homeowner shifts his large electricity consumption to a period of low demand he capitalizes on paying a low energy bill [13].

The time of the day when energy demand is high is called ON-PEAK and the time of the day when the demand is low is referred to as OFF-PEAK while in-between these periods is the MID-PEAK. There is variation in timing in different seasons, for summer peak time is 3pm–8pm while for winter it is 6am–10am while mid peak is 6am–3pm for summer and changes to 10am–5pm during winter. The best plan is usually to move more electricity usage to off-peak which is more preferable between 11.30pm–2.30am [14].

Table 4.2 shows how price of electricity can be reduced by moving most electricity consumption to off-peak periods by computing ToU pricing for 1000 kWh used per month at different periods with a table showing variation in the seasonal pricing periods.

Pricing for each Time of the day is given below

■ On-Peak 13.197 cent per kWh
■ Mid-Peak 7.572 cent per kWh
■ Off-Peak 4.399 cent per kWh

This will bring to an end the usual fixed monthly charge plan where electricity will be charged based on usage.

Table 4.2 Portland General Electric Time of Use Price Compared with Standard Rate

Change of Usage Period of Electricity	On-Peak (kWh)	Mid-Peak (kWh)	Off-Peak (kWh)	Time of Use (ToU)	Standard Rate
To continuously use	206	408	386	$68.06	$68.50
When 10% of on-peak is shifted to off-peak hours	185	408	407	$66.21	$68.50
When 25% of on-peak is shifted to off-peak hours	154	408	438	$63.48	$68.50
When 35% of on-peak is shifted to off-peak hours	134	408	458	$61.73	$68.50

4.7 Solar Energy

The energy from the sun is certainly enormous and will of course remain forever, the process of collecting the energy for use as electricity varies and it is this type of generated energy that is referred to as Solar energy or Solar electricity. The major ways of utilizing sun radiation to generate electricity are described in the following sections.

4.7.1 Solar Panels (Photovoltaic Modules, PV)

This is the most common method of generating electricity from the sun. It is generated through the use of materials that are semiconductors that easily give out electrons when they absorb heat reaching it from the sun, to produce what is called a Solar Panel. These have been used to power houses, industries, and the largest PV generating plant is the 290 MW plant in Arizona [15].

4.7.2 *Solar Thermal (Concentrated Solar Power, CSP)*

The technology used to generate electricity from the sun is not very popular and will produce far more electricity than the PV thus it cannot be utilized for residential but can be used for big scale use like in utility plants. The largest facility using this technology to generate electricity is the 377 MW plant in California's Desert.

The Concentrated Solar Power (CSP) has three different known technologies which vary only in the means of collecting the energy from the sun.

4.7.2.1 Dish Engine Technology

In this method, power is produced from the use of dish shaped parabolic mirrors where gas is heated in a chamber with collected energy from the sun and the heated gas drives a piston of a generator, thereby producing electricity. For more efficiency the dish is mostly attached to a tracking system that maximizes sun radiation [16].

4.7.2.2 Parabolic Trough

This uses trough shaped mirrors which are long enough to concentrate sun energy to a tube filled with liquid which collects heat as it focuses on the sun thereby heating the liquid in the tube connected to a heat exchange system where water is heated to steam that is used in a steam turbine generator to generate electricity. This operates in a continues recycling process, as the heated fluid transfers its heat, the steam cools and condenses, then the same process is continuously repeated over and over. Heated fluid can also be stored for a long period and then be reused when the sun is down [10, 16].

4.7.2.3 Tower Focal Point Concentrated Solar Power

The Focal Point CSP technology is where a large area of flat computer-controlled mirrors which constantly move to get maximum sun reflection all through the day onto a tower that has a collecting tank on it. Molten salt moves in and out of the tank and as a result it is heated to very high temperature of 537.8°C (1,000°F). The heated fluid is sent to a steam boiler where it uses the inherent heat in the fluid to propel a steam turbine to produce electricity.

4.8 Storage Facilities

The usage of energy from the sun gained significant importance with improved storage (battery technology) such as sodium sulfur, lithium ion, lead-acid, metal-air,

Compressed Air Energy Storage (CAES), pumped hydro, flywheels, fuel cells, and super capacitors [17].

4.8.1 Sodium Sulfur (NAS) Batteries

This is manufactured from a combination of two salts, sodium and sulfur. It has an excellent energy density, very good charge and discharge efficiency, and also is low in price which gives it good advantages for use as a storage facility in a smart grid.

Sodium sulfur batteries are also in use as utility-scaled energy storage in Japan, with hundreds of MW as shown in Figure 4.2. NAS batteries were first hosted as a demonstration project in 1992 and they became available commercially in 2002 [18].

4.8.2 Flywheel Storage Device

This revolution in technology of storage facility is regarded as the most used storage method for smart grid storage infrastructure at both transmission and distribution stages. This could be as a result of its rapid response time in discharging stored electricity, high efficiency or long life, and quite insignificant level of maintenance required. It is cylindrically shaped and has a large rotor contained in a vacuum. The flywheel storage device stores electricity in the form of rotational energy, which can also be referred to as spinning mass.

It can store electricity for a smart grid by drawing energy from it as the rotor rotates at a very high speed, as shown in Figure 4.3. To discharge the electricity, it slows down by switching to generation mode thus returning electricity back to the grid for onward distribution [19].

Figure 4.2   Large sodium sulfur (NAS) [18].

Figure 4.3   Rotating flywheel storing energy mode [19].

4.9 Optimization of Storage Devices

It has been emphasized that it is very important to stabilize the energy in smart grid connection due to the intermittent nature of generated energy from the renewable sources. The best form of storage is eminent, should there be urgent need to balance the grid's energy gap. For such optimization, electricity storage with rapid response time is very important and the nature in which it is stored as well as the position of storage device is important. Electricity can be stored in various forms, for example, electrically, mechanically or electrochemically. The storage device has different applicable usage in the grids stabilization, for instance sodium sulfur batteries are very reliable and are used in centralized energy storage systems, pumped hydro storage and CAES systems are used in smart grid for load leveling applications and transmission services while Flywheel batteries are used for power quality applications. The battery energy storage system helps significantly in stabilizing smart grid with a larger quantity of electricity from renewable source by storing the energy when in excess and returning it to the grid when required [20].

Batteries are classified into mobile and stationary batteries. The stationary batteries take surplus electricity from the grid and store it in the form of electrochemical batteries and give back electricity to the grid when there is demand [20].

Mobile batteries are used in the form of EVs batteries which has added advantage in reducing CO_2 and storing electricity when energy is cheap and returning it when expensive. Presently there are two varieties, the EVs and Hybrid Electric Vehicles (HEVs) which can be again divided into six types due to the energy demand as well as power of the batteries [21]. charging the EV could be a problem on managing grid peak but with the advent of decentralization, using multiagent systems will provide solutions to such problems [22].

EVs are classified into two classes while HEVs are classified into four classes [20]. EVs classes are: (1) Battery Electric Vehicles (BEV), and (2) Fuel Cell Electric Vehicle (FCEV). As for the HEVs, they are classified into: (1) Plug-in Hybrid

Electric Vehicle (PHEV), (2) Micro Hybrid Vehicle (MHV), (3) Full Hybrid Electric Vehicle (FHEV), and (4) Mild Hybrid Electric Vehicle (MHEV). Out of all, only two BEVs and PHEVs classifications are suitable for Vehicle-to-Grid (V2G) operation. V2G technology enables the transmission of power between vehicles and the power grid by utilizing the onboard batteries of BEVs and PHEVs as devices for storing energy in the smart grid systems [20].

High optimization can be achieved by optimizing the feet charging load in a power grid with reference to the following factors: (1) Generation capacity, (2) Load characteristics, (3) Power quality, and (4) Grid reliability. Claudio et al. [23] presented and considered some strategies in energy storage which are given as A-posteriori Optimal Strategy algorithm AOS, Dynamic Programming with Markovian request DPM algorithm, Single-Threshold (ST) algorithm, and Dynamic Programming with Independent request (DPI). The strategies need less information on end user power statistics request but provide reduction in price. The following findings were noted, that AOS is certainly not realistic in practical situations, DPM accuracy can be increased by adding other parameters but its optimality is not guaranteed any more but after simulation it highlighted the following:

- ST algorithm in realistic power & battery demand mode is not optimal.
- DPM did achieve noticeable performance gain against DPI.
- DPM could be near AOS upper bound if realistic power is requested.

Therefore, I declared DPM as the best strategy in energy storage [23]. The energy saved during the summertime can be referred to as produced energy [24].

4.10 Connecting Renewable (Solar) Energy to the Smart Grid

As solar technology continues to improve in efficiency and becomes more popular the cost is expected to decrease significantly. Thus, the payback period of a PV system that stands at an average of 20 years could drop to 10 years. This means that even without government incentives it will be cheap to own a PV system which could be cheaper when the system is connected to grid as batteries increase the price due to the fact that they need replacement. Major equipment, referred to as balance-of-system, to integrate with a state utility smart grid or a national smart grid are: (1) Power Conditioning Equipment, (2) Safety Equipment, and (3) Instrumentation and meters. Power Conditioning Equipment practically changes DC to AC as electronics devices use electricity in AC mode. The major power conditioning equipments are:

- Constant DC power to oscillating AC power conversion
- Frequency of the AC cycles should be 60 cycles per second
- Voltage consistency is in the range allowed for output voltage to fluctuate
- Quality of the AC sine curve, regardless of the fact that the AC wave shape is jagged or smooth

Some electric appliances can operate regardless of the electricity quality while others need stabilizers to operate. Inverters are devices required to stabilize the intermittent electricity so that it harmonizes with the requirements of the load in the grid. Safety equipment are devices that provides protection to stand alone or homeowner's system and on-grid integrated electricity generated through a solar or a wind source from being destroyed or endangering people during natural occurrences like storm, lightening, power surges, or a fault that can result from faulty equipment. Instrumentation and meters are equipment that provides a homeowner with the ability to view and manage their electricity in the form of solar energy, the battery voltage of their system, the quantity of electricity one is consuming, and also the strength of the battery in terms of charge/discharge.

4.11 Grid Topology

Depending on the environment as well as the geography of the area a grid is covering, there can be: (1) Radial Distribution, (2) Mesh Distribution, and (3) Looped or Parallel Flow Distribution.

4.11.1 Radial Distribution

This is seen as the cheapest to install and simplest form of topology in grid transmission and distribution which is used generally in sparsely populated localities. It is shaped like a star, showing power distribution from one major supply to the end users. In this topology a single fault can result in a total black out. See for example the traditional grid system, as shown in Figure 4.4.

4.11.2 Meshed Distribution

This is a better topology with more efficient means of transmitting and distributing power. The infrastructure devices are interconnected to allow for cooperation between power sources. This makes power rerouting possible, in an event of a failure between the source and the end user. It is mainly applied in a high loaded area that is quite congested. Figure 4.5 shows the possibility for example of rerouting the power distribution in a congested locality.

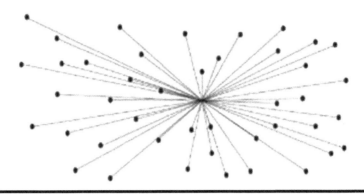

Figure 4.4   Radial topology.

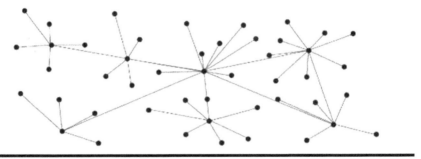

Figure 4.5   Meshed topology.

4.11.3 Looped Distribution

This is the most expensive topology mostly found in European countries including Northern American towns and cities. It is more reliable as it provides the ability of rerouting power in case of fault in a power line before repairs thereby allowing continues service, as represented in Figure 4.6. As suggested by its name it connects two power sources to an area by reconnecting it to the original power source; as a result, an alternative power source is always available.

4.12 Conclusion

The standard of energy provision of any country is universally accepted to be the yard stick in measuring that country's development, be it either underdeveloped, developing, or developed. It is therefore necessary for Nigeria to take the right steps toward bettering the energy availability to its citizens. Presently most Nigerians produces their own electricity which is mostly from fuel generators with few producing from roof-top solar PV. It was reported that, just 37 universities in the

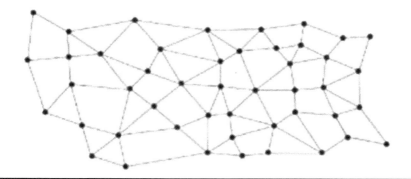

Figure 4.6 Looped topology.

country, use over a thousand generators. Nigeria is rated globally as the highest importer of generators. We believe that, if homeowners can get credit by their local power companies for the electricity produced at their homes through "net-metering" or "feed-in tariff" programs it will encourage more homeowners to participate and eventually increase greatly the energy generation and significantly reduce the greenhouse gas emissions.

The IoT smart grid promises to break these two entry barriers by offering low-cost flexible solutions and using new communication solutions coming from the Internet world, thereby providing more efficient energy production and utilization which will significantly provide the most needed security for the power sector infra-structure against external attack.

References

1. M. I. U. Khan and M. Riaz, "Various Types of Smart Grid Techniques: A review," vol. 7, no. 8 (2016): 7.
2. F. Al-Turjman, "Mobile Couriers' Selection for the Smart-grid in Smart cities' Pervasive Sensing", *Elsevier Future Generation Computer Systems*, vol. 82, no. 1, pp. 327–341, 2018.
3. F. Al-Turjman, "5G-enabled Devices and Smart-Spaces in Social-IoT: An Overview," *Elsevier Future Generation Computer Systems*, 2017. DOI: 10.1016/j.future.2017.11.035
4. F. M. Al-Turjman, H. S. Hassanein, and M. Ibnkahla, "Quantifying Connectivity in Wireless Sensor Networks with Grid-Based Deployments," *Journal of Networkand Computer Applications.*, vol. 36, no. 1 (2013): 368–377.
5. F. Al-Turjman "Modelling Green Femtocells in Smart-grids", *Springer Mobile Networks and Applications*, vol. 23, no. 4, pp. 940–955, 2018.
6. X. Tan, Q. Li, and H. Wang, "Advances and Trends of Energy Storage Technology in Microgrid," *International Journal of Electrical Power &Energy Systems.*, vol. 44, no. 1 (January, 2013): 179–191.
7. U. P. Onochie, H. O. Egware, and T. O. Eyakwanor, "The Nigeria Electric Power Sector (Opportunities and Challenges)," Journal of Multidisciplinary Engineering Science and Technology, vol. 2, no. 4, pp. 494–502, 2015.

8. C. O. A. Awosope, "Nigeria electricity industry: issues, challenges and solutions," *Covenant University's. 38th Public Lecture.*, vol. 3, no. 2 (2014).

9. F. Al-Turjman, "QoS -aware Data Delivery Framework for Safety-inspired Multimedia in Integrated Vehicular-IoT," *Elsevier Computer Communications Journal*, vol. 121, pp. 33–43, 2018.

10. F. Al-Turjman, and S. Alturjman, "Context-sensitive Access in Industrial Internet of Things (IIoT) Healthcare Applications," *IEEE Transactions on Industrial Informatics*, vol. 14, no. 6, pp. 2736–2744, 2018.

11. M. Z. Hasan, and F. Al-Turjman, "Analysis of Cross-layer Design of Quality-of-Service Forward Geographic Wireless Sensor Network Routing Strategies in Green Internet of Things," *IEEE Access Journal*, vol. 6, no. 1, pp. 20371–20389, 2018.

12. O. Patrick, O. Tolulolope, and O. Sunny, "Smart Grid Technology and its Possible Applications to the Nigeria 330 kV Power System," *Smart Grid Renewable Energy*, vol. 4, no. 5 (2013): 391.

13. F. Al-Turjman, and S. Alturjman, "5G/IoT-Enabled UAVs for Multimedia Delivery in Industry-oriented Applications," *Springer's Multimedia Tools and Applications Journal*, 2018. DOI. 10.1007/s11042-018-6288-7

14. G. R. Newsham and B. G. Bowker, "The Effect of Utility Time-Varying Pricing and Load Control Strategies on Residential Summer Peak Electricity Use: A Review," *Energy Policy*, vol. 38, no. 7 (July, 2010): 3289–3296.

15. F. Dinçer, "The Analysis on Photovoltaic Electricity Generation Status, Potential and Policies of the Leading Countries in Solar Energy," *Renewable and* Sustainable Energy Reviews , vol. 15, no. 1 (January,2011): 713–720.

16. D. Mills, "Advances in Solar Thermal Electricity Technology," *Solar Energy*, vol. 76, no. 1 (Januuary, 2004): 19–31.

17. "How Energy Storage Works," *Union of Concerned Scientists*. [Online]. Available: https://www.ucsusa.org/clean-energy/how-energy-storage-works. [Accessed: May 17, 2018].

18. A. Price, S. Bartley, S. Male, and G. Cooley, "A Novel Approach to Utility-Scale Energy Storage," *Power Eng. J.*, vol. 13, no. 3 (1999): 122–129.

19. W. Saad, Z. Han, H. V. Poor, and T. Basar, "Game-Theoretic Methods for the Smart Grid: An Overview of Microgrid Systems, Demand-Side Management, and Smart Grid Communications," *IEEE Signal* Processing Magazine, vol. 29, no. 5 (2012): 86–105.

20. C. Kang and W. Jia, "Transition of Tariff Structure and Distribution Pricing in China," in *IEEE Power and Energy Society General Meeting*, (2011): 1–5.

21. Y. S. Wong, L. L. Lai, S. Gao, and K. T. Chau, "Stationary and Mobile Battery Energy Storage Systems for Smart Grids," *4th International Conference on Electric Utility Deregulation and Restructuring and Power Technologies (DRPT'11)*, Weihai, Shandong, China (2011): 1–6.

22. M. Morte, "E-mobility and Multiagent Systems in Smart Grid," in *17th International Scientific Conference on Electric Power Engineering (EPE'16)*, Prague, Czech Republic (2016): 1–4.

23. F. Al-Turjman, "Energy–aware Data Delivery Framework for Safety-Oriented Mobile IoT," *IEEE Sensors Journal*, vol. 18, no. 1, pp. 470–478, 2017.

24. J. Das and S. Ashok, "Peak Load Management with Wheeling in a Combined Heat and Power Unit under Availability Based Tariff," *Fourth International Conference on Advances in Computing and Communications (ICACC'14)*, (2014): 343–346.

Chapter 5

Smart Meters for the Smart-Cities' Grid

Fadi Al-Turjman* and Mohammad Abujubbeh[†]

Contents

* Antalya Bilim University, Antalya, Turkey
† Middle East Technical University, Ankara, Turkey

5.1 Introduction

Electrical power is considered to be one of the essential factors for the development of societies through improving life quality. However, the conditions in the power industry are changing as electricity demand and renewable integration increase. The increased stress on power demand has produced a burden on the conventional power production resources. With the noticeable decline in conventional power resources reserves and the recent attention on the environmental issues with producing power from fossil fuel based resources, power utilities and investors are motivated to invest in other sustainable ways of power production in order to meet the demand. For instance, one aim of the European 20/20/20 strategy is to increase the share of renewable energy generation up to 20 percent by 2020 [1]. Renewable resources are intermittent by nature. The increased penetration of intermittent renewable distributed generations into the existing power grid makes it challenging for utilities to deliver reliable and good quality power.

With the recent improvements and advancements in technology, there have been efforts to introduce the concept of smart cities [2]. In this context, there will be a noticeable increase in the usage of sensors and sensor networks for providing useful data that enable the efficient control and management of cities [2, 3]. The concept of SCs will not only focus on specific services such as traffic control but will also extend its means to the electric system. With that being said, sensors and sensing networks play an important role for the enforcing the means and measures of a smart grid (SG). In other words, smart meter (SM) usage in SGs ultimately solves most of electrical industry problems [1, 4, 5]. In this way, the SG will be able to effectively deal with many aspects of power generation, transmission, and distribution issues. In addition, it will provide better options for monitoring the status of power delivered to consumers relative to conventional methods.

SMs provide a powerful way of enhancing the power transaction process between the source and sink in a SG. The functionalities of SMs in SGs vary depending on the application objective such as energy demand saving, feedback to consumers, dynamic pricing, and appliances control, depending on demand curves, security enhancement, fault/outage detection, supply quality assessment, etc. [6]. Fault detection and supply performance assessment are two applications that enhance power reliability and quality in a SG.

This study aims to provide a comprehensive review on the role of SMs in SGs with an extensive focus on the power quality (PQ) and power reliability (PR) applications. Another aim is to compare the presented works in literature considering different metrics. The chapter is organized as follows: Section 5.2 reveals the review papers presented in literature that are associated with the usage of SMs in SGs, the third section provides an overview on the key enablers for achieving a successful advanced metering infrastructure (AMI) for SGs, Section 5.4 and 5.5 compare SM related literature considering different metrics in the domain of PQ and PR respectively, Section 5.6 reveals the open research gaps for directing future research, and the last section concludes the thoughts introduced in the chapter with some possible future work suggestions. The following table provides the definitions of acronyms used in this chapter.

5.2 Overview of Related Surveys

The ample work presented in literature discusses the usage of SMs in a SG to enhance the process of power transaction between the supply and the demand sides considering different aspects. In the light of SGs, [7], authors review selected SMs' functionalities with an intense focus on data analysis aspects such as complexity, collection speed, and volume of data. Applications and techniques used in the data analytics process such as Self Organizing Map (SOM), Support Vector Machine (SVM), and Fuzzy Logic (FL) were surveyed. The AMI technology is thoroughly discussed in [6, 8, 9] with a comprehensive review on the functionalities and application benefits of integrating the use of AMI in SGs. References [10, 11] are used to overview the use of AMI technique to enhance employing the technological advancement in residential electricity management schemes. Furthermore, communication techniques and Named Data Networking (NDN) aspects in micro grids have been reviewed in [12]. The NDN-fog computation approach has been also proposed to push data cloud to the edge for the objective of minimizing the latency where it is virtual in various micro grid applications. Among the applications, power outage detections or allocation and voltage and frequency fluctuations monitoring are essential not only for power utilities but also for customer benefits.

Power reliability and quality monitoring provide clear details about the health of a power system. Statistics show that power faults or component failures in a power system cause more than 80 percent of power outages/cuts in a power distribution grid [13, 14]. Hence, researchers must develop new methods and techniques in grid improvement. With the recent advancement in technology and the appearance of SMs and SGs concepts, several academic works have been proposed in this area. The aim of this chapter is to review and assess the effectiveness of these works considering different metrics in the following chapters.

5.3 Advanced Metering Infrastructure Technology

With the recent noticeable shift toward new methods of power generation, distribution, and transmission utilities face challenges that the conventional setup of power systems cannot provide ways and methods for handling. In the conventional power system, it is difficult to have complete information on the power flow in many aspects such as power quality, reliability, energy usage at different loads etc. Recently, there has been a significant move toward renewable ways of energy production which is known for its supply intermittency and the continuous increase in power demand. This, as well as the environmental issues in the conventional energy sources are all considered to be challenges existing in the conventional power system. With that being said, new technologies can be employed to enhance the functionality of power systems to be sufficiently modernized to cope with the continuous changes and challenges which ultimately contributes to the concept of a SG. AMI offers a sustainable solution in this regard which provides a two-way communication scheme between utilities and loads (consumers) as shown in Figure 5.1.

Data including voltage and current readings as well as demand curves will be collected from loads using AMI devices such as SMs, then the data is transferred using AMI to clouds and then to utilities in order to process the data and manage transmission and distribution. Then feedback will be sent to consumers for monitoring their consumption or the quality of the received power.

5.3.1 Smart Meter Internal Structure

A key enabler device in the AMI is SM where it is installed on the consumer side for collecting real-time voltage and current data. Unlike conventional Automatic

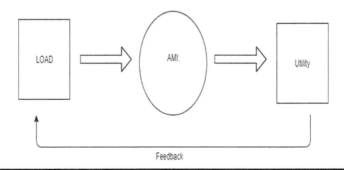

Figure 5.1   Review on the AMI process in SGs.

Meter Reading (AMR) where data collection is monthly, SMs provide the ability of daily data collection [8] via communication networks. Hence, SMs in SGs are beneficial not only for consumers, but also for utilities and the environment. The major features of SMs, but not all features, are [15, 16]

- Energy billing
- Electricity consumption reduction
- Consumption curves for both ends
- Net metering
- Power reliability monitoring: Outage detection
- Power quality monitoring: Harmonics and voltage disturbances classification
- Power security monitoring: Fraud and theft detection
- Automated remote control abilities
- Remote appliance control
- Interfacing other devices
- Indirect greenhouse gases reduction as a result of reduced demand
- Less utility trucks in the streets for outage allocation and PQ tests.

The aforementioned list implies that AMI is able to deal with most of conventional power systems' challenges relative to the AMR technology. It can be said that it is expected to have more complexities in the structure of SMs since it requires integration of high-tech components to provide the good functionalities and features as shown in Figure 5.2.

SMs mainly consist of a microcontroller unit (MCU), a power supply unit with a complimentary battery, voltage and current sensors for active and reactive energy measurement in the energy metering IC, a real-time clock (RTC), and a communication facility as listed below [17]:

5.3.1.1 Microcontroller

MCU is considered to be the heart of a SM where most of the major data processing occurs. Therefore, all operations and functions in the SM are controlled by the MCU including the following:

- Communication with the energy metering IC
- Calculations based on the data received
- Display of electrical parameters, tariff, and cost of electricity
- Smartcard reading
- Tamper detection
- Data management with EEPROM
- Communication with other communication devices
- Power management.

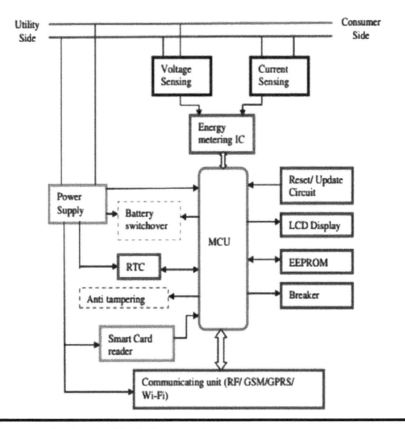

Figure 5.2   The interior block diagram of a SM [18].

Nowadays, most of SMs are equipped with LCD interfaces that enable the consumers not only to learn their electricity tariffs and energy consumption patterns but also to learn the quality of power delivered from utilities as well as the indication of power outages when it occurs. Such functionalities are also processed by the MCU unit.

5.3.1.2 Power Supply Unit

The SM circuit is supplied with power from the main AC lines through AC-DC converters and voltage regulators. A supplemental switchover battery is charged from the main AC lines in order to power the circuit when the connection between main AC and power supply unit is interrupted or a power outage occurs. Solar cells and rechargeable batteries can also be used to supply SM with power during the day [19].

5.3.1.3 Energy Measurement Unit

Based on the voltage and current readings sensed by the voltage and current sensors, energy measurement units perform signal conditioning, and computation of active, reactive, and apparent powers. Energy measurement units can operate as an embedded chip into the MCU or as a standard separated chip to provide the measurements as voltage or frequency pulses.

5.3.2 Wireless Communication in AMI

Data communication in AMI is an essential part where data are instantly collected, transferred to the utility to further process it, and then feedback will be given to the consumers accordingly. In SGs communication areas can be divided into three main categories. The first is Local Area Network (LAN) which describe the communication scheme between consumers and SMs. Secondly, the Neighborhood Area Network (NAN) basically represents a communication medium that contains flowgates to perform specific processes (such as data aggregation, encoding) on the data coming from SMs before it is transmitted to the cloud. The third stage consists of a Wide Area Network (WAN) which is responsible for communicating data between the cloud in a specific region with the destination (utility). The schematic is depicted in Figure 5.3.

However, there are specific areas that need further considerations in AMI communication. In AMI infrastructure, big data transmission, data security,

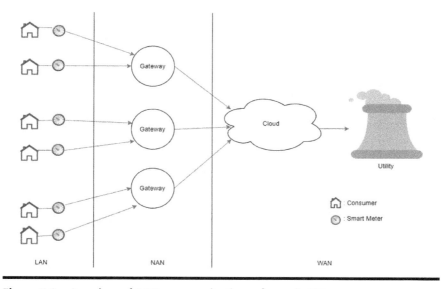

Figure 5.3   A review of AMI communication scheme in SGs.

network scalability, and cost effectiveness are among the essential areas that need more attention [15]. Hence, there is a need for international standards and regulations to put a framework on communication aspects in AMI. In this regard there are various standards developed by international institutions such as IEEE 802.15.4, IEC 61970, ISO 1802 [20] in order to insure reliable, secure, and efficient power delivery to the consumers. Following an analysis on literature, there have been variety of communication protocols developed [21, 22] according to international standards that can be used in LAN, NAN, and WAN regions of the AMI in SGs. The most common used communication protocols in AMI are listed in Section 5.3.2.1.

5.3.2.1 Local Area Network (LAN)

5.3.2.1.1 ZigBee

ZigBee is a communication technology developed according to the IEEE 802.15.4 standard that transmits data at a rate of approximately 250 Kbps on 2.4 GHz frequency [23]. It is well known that ZigBee targets applications that require short domain communication with a range of 0–100 m. Despite the short range of coverage, ZigBee proves itself in the low power consumption and system scalability but at the cost of data transmission rate. Hence, this technology can perform well in the LAN region of a SG.

5.3.2.1.2 Wi-Fi

Wireless Fidelity (Wi-Fi) is another communication technology that is largely used nowadays in homes and business areas. It is designed according to IEEE 802.11b/g/n standard with a capability of data transmission at frequencies 2.4 and 5 GHz in the domain of 0–250 m at a rate of 54 Mbps [24]. The domain of Wi-Fi is larger compared to ZigBee as well as this It also provides a relatively higher rate of data transmission at the cost of scalability. The security in this technology is also an advantage since it has a robust authentication procedure.

5.3.2.1.3 Bluetooth

Bluetooth technology can also be categorized under the short range communication schemes (0–100 m) that is developed according to IEEE 802.15.1 standard to transfer data at a rate of 721 kbps [25] and a frequency of 2.4 GHz. The technology is distinguished by its capability of low power consumption since it has a limited capability of big data transfer. The widespread of the technology among users especially in smartphones gives it another advantage in being integrated in SG LANs.

5.3.2.1.4 Z-Wave

The Z-Wave technology is a radio frequency based communication mainly designed for in-home appliances remote control which relies on an unlicensed 900 MHz frequency which transmits data at a rate reaching 40 kbps for a range of 0–30 m [26]. The operating frequency of this technology reduces the risk of being disrupted by the previously mentioned technologies since they operate on 2.4 GHz frequency which enhances the reliability of this technology.

5.3.2.2 Neighborhood and Wide Area Networks (NAN and WAN)

5.3.2.2.1 Cellular

The cellular technology plays an important role in the AMI of SG. 2G, 3G, and 4G cellular modes that provide the capability of big data transmission that rates from 14.4 kbps (for 2G) until 100 Mbps (for 4G) at a licensed frequency band (824 MHz and 1900 MHz) [27]. The technology targets a large area because of its long domain that ranges between 10–100 km. With that being said, cellular technology consumes high power for the transmission process.

5.3.2.2.2 WiMAX

This communication technology is developed according to IEEE 802.16 standard in which it operates at two frequency bands (2–11 and 11–66 GHz) to transmit data at rate of 70 Mbps for an area domain of approximately 50 km [28]. The WiMAX has a similar principle to Wi-Fi but differs in the distance of transmission.

5.3.2.2.3 Sigfox

Sigfox is a developing machine to machine WAN communication solution that operates on a frequency band of 868 MHz that has the ability to cover 30–50 km in rural areas and 3–10 km in urban areas to deliver the data at a rate of 100 bps [29]. An advantage with this technology is noticeable in its low power consumption for data transmission. The limitation of Wi-Fi technology to be applied in NAN or WAN is the short area coverage while the cellular provided a solution for big data transmission in WANs at the cost of power consumption, Sigfox seems to provide a middle-way solution considering coverage and power consumption.

5.3.2.2.4 LoRaWAN

LoRaWAN is a recent non-profit organization that is mainly found to be integrated in Internet of Things (IoT) WAN applications. The technology operates on a 900 MHz frequency to transmit data at a rate of 50 kbps for distance ranges between 10–15 km in rural areas and 2–5 km in urban areas [30]. Relative to Sigfox technology considering the trade-off between coverage and power consumption, LoRaWAN offers a better data rate with reduced power consumption. Table 5.1 illustrates a summary of communication technologies used in LAN, NAN, and WAN areas of a SG.

5.3.3 Routing Algorithms

One of the essential SG characteristics that differ from conventional power grids is the communication network. The new setup of communication networks in SGs should be distinguished by their ability to support time-sensitive and data-intensive management tasks [31]. For instance, the closer SMs are to the gateways, the more information they will transfer which leads to more data concentration which may affect the transmission reliability. Hence, choosing the appropriate routing algorithm is an essential area to be covered in SG communication network setups [32]. Data routing is moving one data unit through one possible communication path or more from the source to the destination node [33]. Having multiple paths to transmit the data unit from the source to the destination increases the network robustness relative to one single path in which there is a higher possibility for the path to fail. Therefore, a robust routing scheme implies adding more data transmission complexity. Since there may be multiple paths for data transmission in SG communication networks, a routing algorithm decides on which path to be used for the data transmission considering different metrics. It can be said that reliability, cost, computational power, delay, and data throughput are some of the key objectives when developing a routing algorithm [34] for SG applications. Routing algorithm design objectives as revealed from literature are listed in Section 5.3.3.1.

Table 5.1   Acronyms of Definitions Used in this Chapter

Acronym	Definition
SG	Smart Grid
SM	Smart Meter
AMI	Advanced
AMR	Automatic Meter Reading
PQ	Power Quality
PR	Power Reliability

5.3.3.1 Delay

In SG communication networks if the transmitted information from the source to the destination is received late, undesired events might happen even though the data control devices take a correct action [35]. Therefore limiting, if not eliminating, delays in the communication network is essential to enhance the quality of delivering data collected. Route selection is the essence of limiting the delay in SG communication networks. Authors in [35] estimate the delay and propose the minimum end-to-end delay multicast tree routing algorithm to select the path with least delay in NAN region of SGs. Another routing algorithm is developed in [36] to ensure an efficient communication path selection for SG applications in the NAN communication region. The developed Dijkstra based Dynamic Neighborhood Routing Path Selection (DNRPS) algorithm shows its superiority in the reduced communication delay by a highest gain percentage reduction of 19 percent over other selected algorithms and Dijkstra is one of them. Delay is also studied in [37] where authors employ the Artificial Cobweb Routing Algorithm (ACRA) to enhance the Quality of Service (QoS) of the network by focusing on the delay and data throughput for applications in low voltage power distribution networks, in this context, the LAN region. Moreover, study [38] proposes routing algorithm for QoS enhancement considering delay, memory utilization, packet delivery ratio (PDR), and throughput using the energy efficient and QoS-aware routing protocol (EQRP) algorithm that is inspired by Bird Mating Optimization (BMO). Furthermore, authors in [39] attempt to study the delay for SG applications using the Greedy QoS Routing algorithm. Whereas in reference [40] the Straight-Line Path Routing (SLPR) algorithm is used. Delay in some works [41–43] is referred to as latency. Reference [41] employs the layered cooperative processing algorithm to enhance the QoS based on latency and reliability for SG application. In reference [42] the latency and reliability are objectives enhanced using an improved Ant Colony Optimization algorithm for the application in NAN region. Moreover, in reference [43] authors used the Greedy Backpressure Routing algorithm to study latency as well as throughput. Following the analysis on this literature, various algorithms are used and the application of each study differs such that in some works the region of application in the SG (LAN, NAN, or WAN) is targeted and some works target the SG in general as illustrated in Table 5.2.

5.3.3.2 Lifetime

SMs in the communication network of a SG are scalable, flexible, and intelligent nodes. However, these sensor nodes are subject to failure due to their limited energy capacities which shorten the overall network lifetime. For this, authors in [44] propose an energy routing based Wireless Sensor Network (WSN) scheme to lengthen the network lifetime considering the sensor node energy limits for applications in power distribution networks whereas in [45] authors show the advantage of using

Table 5.2 Categorization of Communication Technologies in SG AMI

Application in SGs	Technology	Data Rate	Coverage	Frequency	Standard
LAN	ZigBee	250 kbps	0–100 m	2.4 GHz	IEEE 802.15.4
	Wi-Fi	54 Mbps	0–250 m	2.4 and 5 GHz	IEEE 802.11b/g/n
	Bluetooth	721 kbps	0–100 m	2.4 GHz	IEEE 802.15.1
	Z-wave	40 kbps	0–30 m	900 MHz	ITU-T G. 9959
NAN & WAN	Cellular	14.4 kbps (2G) 100 Mbps (4G)	10–100 km	824 MHz and 1900 MHz	GSM/ GPRS/EDGE, (2G), UMTS/HSPA (3G), LTE (4G)
	WiMAX	70 Mbps	0–50 km	2–11 & 11–66 GHz	IEEE 802.16
	Sigfox	100 bpts	30–50 km (Rural) 3–10 km (Urban)	868 MHz	Sigfox
	LoRaWAN	50 kbps	10–15 km (Rural) 2–5 km (Urban)	900 MHz	LoRaWAN

Maximum Likelihood Routing Algorithm (MLRA) over a random method for extending the lifetime of a SG wireless network.

5.3.3.3 Cost

Routing cost is another important objective that should be considered when designing a routing algorithm. It is proposed in [46] that relying on the Open Shortest Path First (OSPF) technique can help reduce the cost of route selection for applications in LANs of SGs. Unlike reference [47] where authors rely on FL, namely Neurofuzzy-based Optimization Multi-constrained Routing (NFMCR) algorithm, for cost and error reduction in route selection for application in SGs. Cost alongside fault tolerance is also studied in reference [48] where authors employ the Dijkstra shortest path routing algorithm considering all SG communication regions, LAN, NAN, and WAN. In another study [49], authors also employ the Dijkstra's shortest path routing algorithm to find the least routing cost among the network nodes putting the security of the grid into consideration.

5.3.3.4 Reliability

Reliability in communication networks has something to do with failures when transmitting information from the source to the destination [50]. From the hardware point of view, reliability can be enhanced through the inclusion of more links and components to overtake the work of faulty components in the failure situation [51] which increases the complexity and installation costs. Hence it can be said that a reliable communication network is fault/failure and error tolerant which will be referred to as Error and Fault Tolerance (FET). Some studies attempt to improve network reliability considering error reduction [47], fault tolerance [48], or failure probability [52]. Some other studies also examine the PDR as an indicator of reliable communication. PDR is the ratio between the amount of data packets sent by a source and data packets received at the destination [53]. In SG domain, PDR is improved in study [38] using EQRP as mentioned previously. Moreover, the reliability of the link between two nodes in reference [41] is assessed using PDR and a reliability routing decision algorithm is used to select the reliable routing path. Similar to PRD, In reference [42] authors attempt to reduce packet-loss rate (PLR) using an improved ant colony optimization for applications in SG NAN regions. Furthermore, both PDR and PLR are considered in [54] for applications in NAN using a Hybrid Metric algorithm. Based on literature, it can be said that reliability enhancement is essential and should be studied in SG LAN networks since data congestion is higher on SMs around gateways.

5.3.3.5 Throughput

Unlike PDR, throughput has something to do with the total number of data packets that are successfully transmitted to the destination over the time [53] which can be interpreted as number of bits processed over time. The throughput is considered as a secondary objective in studies [32, 38, 43, 54] using various algorithms and application regions. For instance study [32] provides a routing algorithm for a generic application in SGs whereas in [43] authors propose their algorithm to enhance the throughput in NAN region specifically for data aggregation points, in this context, gateways. Moreover, study [55] develops the transmission time for remaining path (TTRP) and metric-based opportunistic routing (TTRPOR) to enhance the LAN wired communication data throughput. In light of SG more research should be undertaken on the throughput enhancement for the three regions of communication (LAN, NAN, and WAN) in SGs.

5.4 Smart Meters and Power Quality

Power quality simply means delivering a smooth and a steady voltage waveform to consumers. PQ issues are majorly found in variant voltage waveforms and supply frequency. These waveform disturbances accompanied with current and voltage distortions can be caused by changeable load patterns [56]. Hence it is essential for utilities and consumers to detect and classify those disturbances as it will enhance the power grid performance. Using SMs in assessing PQ is considered to be an important switching point as it implies the application of SGs measures. The employment of SMs into PQ assessment is observed in literature works and hence introduced in the following subsections.

5.4.1 Assessment Parameters

The global importance of PQ assessment in power distribution networks [57–59] brought great attention to standardizing PQ guidelines in order to promote its assessment accuracy. Many organizations successfully introduced international standards for PQ assessment. The Institute of Electrical and Electronics Engineers (IEEE) [60], and the International Electro-technical Commission (IEC) [61] are among the reputable organizations that significantly contributed in developing PQ standards. For instance, IEEE 519-1992 is an IEEE recommended practice for harmonics control in electrical power systems that intendeds to provide steady-state operation limits in order to minimize harmonics and transients [62]. This practice is widely adopted by North American power utilities [63]. Subsequently, the IEEE 1159–1995 emerged aiming to build a guideline for acceptable methods of monitoring PQ in power distribution networks [64]. In addition, it classifies the typical characteristics of electromagnetic phenomena parameters that mainly causes PQ. Two standards (IEC 61000-4-7 and IEC 61000-4-15) developed by the IEC already exist [65] in which they included power quality parameters along with their calculations and interpretation methods. Further developments continued to introduce the IEC 61000-4-30 standard [66]. Furthermore, EN50160 is a European power quality standard that is adopted by many European countries [67]. It is essential to consider standardized parameters while assessing PQ. The common parameters introduced by international standards will be used for the comparison to draw a clearer picture of what mostly causes poor power quality.

Technological advancements are essential for PQ assessment [68] as it needs capable devices and instruments and algorithms to achieve accurate monitoring that meets the required standards and regulations [68, 69]. AMI technology plays an important role in this regard. Morales-Velazquez et al. and Koponen et al. attempted to develop a SM network for electrical installation monitoring for PQ assessment parameters including power factor, Total Harmonic Distortion (THD), and voltage dips, swells, and interruptions[68, 70]. Authors consider an in situ big data processing capability relying on a Field-Programmable Gate Array (FPGA) embedded in the SM. Unlike Koziy et al. and McBee et al., where they introduce a PQ assessment SM putting cost of measurement device into consideration [69, 71]. They use voltage transient detection, current drop patterns, and arc-fault detection as their evaluation parameters. In Koponen et al. and Borges et al. voltage distortion and imbalance were used to evaluate PQ and provide an advanced warning of PQ problems as well as free data processing capability [70, 72]. Furthermore, a smart monitoring system is proposed in McBee et al. and Morsi that relies on voltage deficiency indications to reduce customer complaints and operational costs [71, 73]. In Borges et al. and Capua and Romeo the authors consider reduction of computational effort for PQ disturbances classification such as sags, swells, flickers, Harmonic Distortion (HD), voltage interruptions, and oscillatory transients [72, 74]. That is to say, studies can be categorized according to the parameter(s) used in their assessment criteria as shown in Table 5.3.

5.4.2 Techniques Embedded in SM for PQ Analysis

PQ Disturbances' classification and detection has been an important topic that many researchers continuously attempt to solve [72, 74]. Various methods and techniques have been introduced in literature to efficiently provide the status of disturbances in a power system. Regardless of using SMs in power quality assessment, the commonly used techniques are based on either, signal processing techniques (Fast Fourier Transform (FFT), Wavelet Transform (WT), S-Transform (ST), etc.), or artificial intelligence approaches (Artificial Neural Networks (ANN), FL, and SVM), or heuristic optimization approaches (Particle Swarm Optimization (PSO), Genetic Algorithms (GA), etc.) [72, 74–78]. However, these techniques and studies presented literature focus on PQ assessment in the context of a conventional power grid. In other words, with the emergence of SGs definitions and measures accompanied with the numerous deployments of SMs, PQ analysis techniques should shift toward smart ways of implementation. A key factor of achieving the smartness of PQ analysis can be introducing in SMs-embedded analysis techniques. It is worth mentioning that this will enable having SMs with multi-functionalities alongside PQ assessment such as load profile monitoring, energy billing, outage detection, or even remote automated switch control in SGs. In this context, based on the limited literature related two techniques are revealed and compared in Section 5.4.2.1.

5.4.2.1 Wavelet Transform (WT)

WT is mathematical model that plays an important role in signal analysis for the purpose of PQ assessment as it provides the ability to analyze waveforms characteristics in time-frequency domain [76, 78]. The WT is useful when voltage transients as well as short duration voltage variations (sags, swells, interruptions) are considered in the PQ studies [77, 79]. In the light of techniques embedded in SMs for PQ studies, it is worth mentioning selected studies. For instance, in Morsi and Granados-Liberman et al. the authors employ the Discrete WT (DWT) to measure reactive energy with the presence of time variant PQ disturbances like voltage swells and harmonics for the objective of maintaining a less computational effort relative to Wavelet Packet Transform (WPT) [73, 78]. Whereas in Koziy et al. and McBee and Simoes a wavelet multiresolution (WMT) approach is used to detect voltage transients and current drops as well as a THD measurement Goertzel algorithm considering low smart meter cost [69, 71]. A developed technique relying on employing

Table 5.3 Studies Categorization According to Used Assessment Parameters

Ref.	Transients	Sags	Swells	Under/over Voltage	Voltage Inbalance	Voltage Interruptions	Dips	Power Frequency	Flicker	Reactive Power	Harmonics
[68]			✓✓		✓	✓✓	✓	✓			✓
[69]	✓		✓			✓		✓			✓
[70]				✓	✓						✓
[71]				✓							
[72]	✓	✓	✓			✓			✓		✓
[73]										✓	
[74]											

the Recursive Pyramid Algorithm (RPA) in applying DWT was introduced in Chen and Zhu, and Hasan et al., where the primary aim is showing the computational efficiency of the proposed model over the normal DWT [79–80]. Following an analysis presented in Tse et al., a novel integer lifting WT (ILWT) technique is employed with SMs to achieve real-time compression and transmission of signals for the purpose of analyzing harmonics and PQ short duration disturbances such as sags, swells, voltage transients and interruptions, and flickers [81].

Following the analysis of the literature, it can be said that the main advantage of using WT is observed in its ability to provide a good analysis resolution in the time-frequency domain and hence short duration voltage variations can be classified as mentioned previously. However, WT can induce computational burden on the data processing units in SMs (though the DWT is more efficient in terms of computation effort as shown in Table 5.4) especially when better analysis performance is desired, which is a major disadvantage of this technique.

5.4.2.2 Fast Fourier Transform (FFT)

FFT is another widely used signal analysis technique which mainly converts signals from time to frequency domain [82]. The usefulness of this transform is observed in periodical signals, namely, in identifying their phases and amplitudes [83] and hence determining noticeable harmonic events. As in Jia et al., it is used to achieve a low computational burden on the hardware while evaluating the HD [32]. In this study, the model was only applied on harmonics extraction and two other artificial intelligent approaches (ANN and decision trees) are used to extract short term PQ disturbances including sags, swells, and oscillatory transients. The simplicity of implementation in FFT makes it a noticeable fast processing technique relative to other complex techniques. In Junput et al., FFT is preferred for its fast performance and accuracy to detect and estimate THD when embedded in a SM [84].

One advantage of using FFT in PQ assessment alongside with SMs is that it provides fast performance and accuracy for harmonics evaluation. Another advantage is that FFT is observed to be suitable to be embedded in SMs for PQ assessment compared to other techniques [72, 84]. Nevertheless, it can be mentioned that FFT shows weak performance in terms of short term variations detections and time-frequency domain resolution relative to WT. Table 5.5 shows the usages of both techniques embedded in SMs for different applications and objectives.

5.5 Smart Meters and Power Reliability

Power reliability is related to the total electric interruptions in a power system that has to do with the full loss of voltage waveform unlike power quality which covers voltage sags, swells, and harmonics [85]. A highly reliable power system means that power will be delivered to the consumers all the time without any interruptions. However, power systems are not ideal and several reasons can threaten their reliability which

Table 5.4 Categorization of Routing Algorithms According to Objectives in SG AMI

Ref.	Delay	Lifetime	Cost	Reliability PDR	PLR	FET	Throughput
[35, 36, 39, 40]	✓						
[37, 43]	✓						✓
[38]	✓			✓			✓
[41]	✓			✓			
[42]	✓				✓		
[44, 45]		✓					
[46, 49]			✓				
[47, 48]			✓			✓	✓
[52]						✓	
[54]				✓	✓		
[55]							✓

Table 5.5 Techniques Embedded in SMs for Different Applications and Objectives

Technique		Application(s)	Objective	Ref.
WT	DWT	Reactive Power	Computational effort	[73]
	RPA -DWT	Reactive Power	Computational effort	[69]
	WMT	Voltage transients	Low device cost	[80]
	ILWT	Harmonics and disturbances	–	[81]
FFT		Harmonics	Computational effort	[71]
		Harmonics	Fast performance	[84]

means economic loss to both utilities and consumers. It is estimated that 80 percent of power reliability issues occur in the power distribution network [86]. Hence, in conventional power systems the details of interruptions and outages are hidden. With technological advancements, AMI technology and data analysis techniques are proved to enhance the exposure of power system issues and measure the severity of interruptions [87] which enhances the movement toward the concept of a SG. In the light of SMs, there have been many ways and techniques used in literature to evaluate the reliability of the delivered power from utilities to consumers. This is achieved through determining defined reliability indices. Whereas some other techniques in literature attempt to detect power outages interruptions in a power system, various standards are presented in literature to assess power reliability in a power distribution network. The IEEE1366-2012 standard is a widely used standard that presents that indices can be used in assessing power reliability [88], which presents sustained and momentary interruption indices as well as load-based indices. Frequency, duration, and the extent of the interruption are essential parameters used in characterizing reliability of the power system [89]. Indices presented by IEEE1366-2012 are briefly shown in Table 5.6. Sensors are deployed with sufficient amounts in high and medium voltage networks (HV, MV) which enables the accurate monitor of power reliability events. With the introduction of AMI in low voltage (LV) networks, namely SMs, there is a great opportunity of enhancing the monitoring capabilities on the LV side. That is to say, reliability indices can be accurately calculated with the help of SMs at the LV sides. As done in Kuhi et al., authors present a method of calculating temporospatial disaggregated reliability SAIDI index relying on SM data [87].

In this context, SMs also provide the ability of interruption time reduction relative to conventional meters. Replacement of conventional meters with SMs in a power distribution utility in Brazil is discussed in [90] considering SAIDI index and expected energy not supplied (EENS). The study shows a noticeable annual reduction of both SAIDI and EENS for the period between 2011 and 2015 where they achieved a 16.54 percent reduction in 2015. Unlike a study by Siirto et al., implemented in Helsinki, Finland shows that the help of SMs enabled utilities to achieve a percentage reduction of 50 percent in SAIDindex [89]. AMI technology had more focus in Shang-Wen Luan et al., which develops a ZigBee based automated reliability system that is able to calculate and display reliability indices including sustained and load based indices [91]. Following the focus on employing SMs in calculating reliability indices, Gamroth et al. propose an evaluation model for Power Reliability Index (PRI) which uses the PQ data recorded by the SMs [92]. The usefulness of this study is observed in employing both PQ and reliability considerations to enhance the overall grid system operation with an intense focus on SMs. This is unlike Mohsenzadeh et al., which briefly considers the usage of SMs in this regard but attempts to calculate reliability indices such as ASIDI and ENS as an additional adjective [93]. Following analysis on this limited literature in the domain of using SMs for PR indices calculations, applications vary in the indices calculated as well as the focus on the usage of SMs as summarized in Table 5.7.

Table 5.6   Summary of IEEE1366-2012 Reliability Indices

Index	Description	
Sustained Interruption (SI)	SAIFI	System Average Interruption Frequency Index
	SAIDI	System Average Interruption Duration Index
	CAIDI	Customer Average Interruption Duration Index
	CTAIDI	Customer Total Average Interruption Duration Index
	CAIFI	Customer Average Interruption Frequency Index
	ASAI	Average Service Availability Index
Load based (LB)	ASIFI	Average System Interruption Frequency Index
	ASIDI	Average System Interruption Duration Index
	$CEMI_n$	Customer Experiencing Multiple Interruptions
	CELID	Customer Experiencing Long Interruption Duration
Momentary interruptions (MI)	MAIFI	Momentary Average Interruption Frequency Index
	$MAIFI_E$	Momentary Average Interruption Event Frequency Index
	$CEMSMI_n$	Customer Experiencing Multiple Sustained Interruption and Momentary Interruption Events Index

5.6 Open Research Issues

According to analysis on literature in this chapter, there are some areas that have not been well covered in this field. Firstly, routing algorithms for SG applications considering lifetime has not been fully covered as in Section 5.3, Table 5.4 shows that there was only one reference attempt to design a routing algorithm considering the lifetime objective. Secondly, in Section 5.4, Table 5.3 shows that the study that

Table 5.7   Summary of SM-Based PR Indices Calculation

Ref.	Index(s)	Focus on SMs
[87]	SAIDI	Moderate
[89]	SAIDI	Moderate
[90]	SAIDI and EENS	High
[91]	SI and LB	High
[92]	PRI	High
[93]	ASIDI and ENS	Low

most focused on assessing PQ using SMs is De Capua and Romeo but still needs to be improved in a such a way that contains other assessment parameters such as dips and supplied power frequency [74]. Furthermore, in Section 5.4 Table 5.5, it can be said that there is a need for the development of more techniques and algorithms such as heuristic and artificial intelligence approaches to be embedded into SMs for assessing power quality. Finally, in Section 5.55, it is clearly shown in Table 5.7 that the number of studies employing SMs for calculating MI indices is low.

5.7 Conclusion

The importance of PQ and PR in power grids has brought increased attention to developing new ways and techniques for their assessment. IoT is believed to provide a long term solution to help solving such problems as well as providing the essential building blocks for enhancing the measures of SGs with the help of AMI and SMs technologies. For this study, we reviewed the effectiveness of employing these technologies onto conventional power grids for PQ and PR assessments. The structure of AMI and SM technologies including wireless communication technology as an enabler for IoT as well as data routing algorithms are discussed and open research areas are suggested accordingly.

References

1. P. Kukuča and I. Chrapčiak, "From Smart Metering to Smart Grid," *Measurement Science Review*, vol. 16, no. 3 (2016): (142–148).
2. R. Morello, S. C. Mukhopadhyay, Z. Liu, D. Slomovitz, and S. R. Samantaray, "Advances on Sensing Technologies for Smart Cities and Power Grids: A Review," *IEEE Sensors Journal*, vol. 17, no. 23, 7596–7610, December, 2017.

3. "Home—IEEE Smart Cities." [Online]. Available: https://smartcities.ieee.org/. [Accessed: November, 2017].
4. S. Darby, "Smart Metering: What Potential for Householder Engagement?," *Building Research and Information*, vol. 38, no. 5 (2010): 442–457.
5. S. S. S. R. Depuru, L. Wang, and V. Devabhaktuni, "Smart Meters for Power gGrid: Challenges, Issues, Advantages and Status," *Renewable and Sustainable Energy Reviews*, vol. 15, no. 6 (2011): 2736–2742.
6. G. R. Barai, S. Krishnan, and B. Venkatesh, "Smart Metering and Functionalities of Smart Meters in Smart Grid—a Review," in *2015 IEEE Electrical Power and Energy Conference (EPEC'15)*, London, ON, Canada (2015): 138–145.
7. S. Pawar and B. F. Momin, "Smart Electricity Meter Data Analytics: A Brief Review," in *IEEE Region 10 Symposium (TENSYMP'17)*, Cochin, India (2017): 1–5.
8. S. Nimbargi, S. Mhaisne, S. Nangare, and M. Sinha, "Review on AMI Technology for Smart Meter," in *2016 IEEE International Conference on Advances in Electronics, Communication and Computer Technology (ICAECCT'16)*, Pune, India (2016): 21–27.
9. Q. Sun Hailong Li, Zhanyu Ma, Chao Wang, Javier Campillo, Qi Zhang, Fredrik Wallin, and Jun Guo, "A Comprehensive Review of Smart Energy Meters in Intelligent Energy Networks," *IEEE Internet Things J.*, vol. 3, no. 4 (August, 2016): 464–479.
10. A. Rajabi, L. Li, J. Zhang, J. Zhu, S. Ghavidel, and M. J. Ghadi, "A Review on Clustering of Residential Electricity Customers and its Applications," in *20th International Conference on Electrical Machines and Systems (ICEMS'17)*, Sydney, NSW, Australia (2017): 1–6.
11. A. Rajabi, L. Li, J. Zhang, and J. Zhu, "Aggregation of Small Loads for Demand Response Programs—Implementation and Challenges: A Review," in *2017 IEEE International Conference on Environment and Electrical Engineering and 2017 IEEE Industrial and Commercial Power Systems Europe (EEEIC/I&CPS Europe)*, Milan, Italy (2017): 1–6.
12. K. Monteiro, M. Marot, and H. Ibn-khedher, "Review on Microgrid Communications Solutions: a Named Data Networking—Fog Approach," in *16th Annual Mediterranean Ad Hoc Networking Workshop (Med-Hoc-Net)*, (2017): 1–8.
13. Y. Jiang, C.-C. Liu, M. Diedesch, E. Lee, and A. K. Srivastava, "Outage Management of Distribution Systems Incorporating Information from SmartMmeters," *IEEE Trans. Power Syst.*, vol. 31, no. 5 (September, 2016): 4144–4154.
14. A. A. Chowdhury and D. O. Koval, "Reliability principles," in *Power Distribution System Reliability*, Hoboken, NJ, USA: John Wiley & Sons, Inc., 45–77, n.d.
15. R. Rashed Mohassel, A. Fung, F. Mohammadi, and K. Raahemifar, "A Survey on Advanced Metering Infrastructure, *International Journal of Electrical Power & Energy Systems*, vol. 63 (December, 2014): 473–484.
16. F. Al-Turjman, "Modelling Green Femtocells in Smart-grids," *Springer Mobile Networks and Applications*, vol. 23, no. 4, pp. 940–955, 2018.
17. K. S. Weranga, S. Kumarawadu, and D. P. Chandima, *Smart Metering Design and Applications*. Singapore: Springer Singapore, 2014.
18. A. K. Chakraborty and N. Sharma, "Advanced Metering Infrastructure: Technology and Challenges," in *2016 IEEE/PES Transmission and Distribution Conference and Exposition (T&D)*, (2016): 1–5.
19. G. A. Ajenikoko and A. A. Olaomi, "Hardware Design of a Smart Meter," *J. Eng. Res. Appl.*, ISSN : 2248–9622, Vol. 4, Issue 9(Version 6), September 2014, pp.115–119.

20. D. Baimel, S. Tapuchi, and N. Baimel, "Smart grid communication technologies—overview, research challenges and opportunities," in *International Symposium on Power Electronics, Electrical Drives, Automation and Motion (SPEEDAM'16)*, Anacapri, Italy (2016): 116–120.

21. N. Saputro, K. Akkaya, and S. Uludag, "A Survey of Routing Protocols for Smart Grid Communications," *Comput. Networks*, vol. 56, no. 11 (July, 2012): 2742–2771.

22. L. Chhaya, P. Sharma, A. Kumar, and G. Bhagwatikar, "Communication Theories and Protocols for Smart Grid Hierarchical Network," *J. Electr. Electron. Eng.*, vol. 10, no. 1 (2017): 43–48.

23. A. Y. Mulla, J. J. Baviskar, F. S. Kazi, and S. R. Wagh, "Implementation of ZigBee/802.15.4 in Smart Grid Communication and Analysis of Power Consumption: A Case Study," in *2014 Annual IEEE India Conference (INDICON'14)*, India, (2014): 1–7.

24. F. Al-Turjman, "Mobile Couriers' Selection for the Smart-grid in Smart cities' Pervasive Sensing," *Elsevier Future Generation Computer Systems*, vol. 82, no. 1, pp. 327–341, 2018.

25. N. Shaukat et al., "A Survey on Consumers Empowerment, Communication Technologies, and Renewable Generation Penetration within Smart Grid," *Renewable and Sustainable Energy Reviews*, vol. 81, (January, 2018): 1453–1475.

26. F. Al-Turjman, "Energy-aware Data Delivery Framework for Safety-Oriented Mobile IoT," *IEEE Sensors Journal*, vol. 18, no. 1, pp. 470–478, 2017.

27. A. Mahmood, N. Javaid, and S. Razzaq, "A Review of Wireless Communications for Smart Grid," *Renewable and Sustainable Energy Reviews*, vol. 41, (January, 2015): 248–260.

28. A. Usman and S. H. Shami, "Evolution of Communication Technologies for Smart Grid Applications," *Renewable and Sustainable Energy Reviews*, vol. 19, (March, 2013): 191–199.

29. F. Al-Turjman, "QoS-aware Data Delivery Framework for Safety-inspired Multimedia in Integrated Vehicular-IoT," *Elsevier Computer Communications Journal*, vol. 121, pp. 33–43, 2018.

30. H. G. S. Filho, J. P. Filho, and V. L. Moreli, "The Adequacy of LoRaWAN on Smart Grids: A Comparison with RF Mesh Technology," in *2016 IEEE International Smart Cities Conference (ISC2'16)*, Trento, Italy (2016): 1–6.

31. W. Wang, Y. Xu, and M. Khanna, "A Survey on the Communication Architectures in Smart Grid," *Comput. Networks*, vol. 55, no. 15 (October, 2011): 3604–3629.

32. X. Jia, X. Chen, S. Shao, and F. Qi, "Routing Algorithm of Smart Grid Data Collection Based on Data Balance Measurement Model," in *The 16th Asia-Pacific Network Operations and Management Symposium*, Hsinchu, Taiwan (2014): 1–4.

33. Wen-Ning Hsieh and Gitman, "Routing Strategies in Computer Networks," *Computer (Long. Beach. Calif.)*, vol. 17, no. 6 (June, 1984): 46–56.

34. P. Bell and K. Jabbour, "Review of Point-to-Point Network Routing Algorithms," *IEEE Communications Magazine*, vol. 24, no. 1, (January, 1986): 34–38.

35. R. Hou, C. Wang, Q. Zhu, and J. Li, "Interference-Aware QoS Multicast Routing for Smart Grid," *Ad Hoc Networks*, vol. 22 (November, 2014): 13–26.

36. Q. Wang and F. Granelli, "An Improved Routing Algorithm for Wireless Path Selection for the Smart Grid Distribution Network," in *2014 IEEE International Energy Conference (ENERGYCON)*, Cavtat, Croatia (2014): 800–804.

37. L. Zhang, X. Liu, Y. Zhou, and D. Xu, "A Novel Routing Algorithm for Power Line Communication over a Low-Voltage Distribution Network in a Smart Grid," *Energies*, vol. 6, no. 3 (March, 2013): 1421–1438.

38. F. Al-Turjman, "Information-Centric Framework for the Internet of Things (IoT): Traffic Modelling & Optimization," *Elsevier Future Generation Computer Systems*, vol. 80, no. 1, pp. 63–75, 2017.

39. F. Al-Turjman, "QoS -aware Data Delivery Framework for Safety-inspired Multimedia in Integrated Vehicular-IoT," *Elsevier Computer Communications Journal*, vol. 121, pp. 33–43, 2018.

40. A. Noorwali, R. Rao, and A. Shami, "End-to-end delay analysis of Wireless Mesh Backbone Network in a Smart Grid," in *IEEE Canadian Conference on Electrical and Computer Engineering (CCECE'16)*, Vancouver, BC, Canada (2016): 1–6.

41. W. Sun, J. P. Wang, J. L. Wang, Q. Y. Li, and D. M. Mu, "QoS Routing Algorithm of WSN for Smart Distribution Grid," Advanced *Material Research*, vol. 1079–1080, (December, 2014): 724–729.

42. M. Z. Hasan, and F. Al-Turjman, "Optimizing Multipath Routing With Guaranteed Fault Tolerance in Internet of Things," *IEEE Sensors Journal*, vol. 17, no. 19, 6463–6473, 2017.

43. Bin Hu and H. Gharavi, "Greedy Backpressure Routing for Smart Grid Sensor Networks," in *IEEE 11th Consumer Communications and Networking Conference (CCNC'14)*, Las Vegas, NV, (2014): 32–37.

44. J. Guo, J. Yao, T. Song, J. Hu, and M. Liu, "A Routing Algorithm to Long Lifetime Network for the Intelligent Power Distribution Network in Smart Grid," in *IEEE Advanced Information Technology, Electronic and Automation Control Conference (IAEAC'15)*, Chongqing, China (2015): (1077–1082).

45. X. Li, Q. Liang, and F. C. Lau, "A Maximum Likelihood Routing Algorithm for Smart Grid Wireless Network," *EURASIP Journal on Wireless Communications and Networking*, vol. 2014, no. 1 (December, 2014): 75.

46. R. Wang, J. Wu, Z. Qian, Z. Lin, and X. He, "A Graph Theory Based Energy Routing Algorithm in Energy Local Area Network," *IEEE Transactions on Industrial Informatics*, vol. 13, no. 6 (December, 2017): 3275–3285.

47. R. Rastgoo and V. S. Naeini, "A Neurofuzzy QoS-Aware Routing Protocol for Smart Grids," in *2014 22nd Iranian Conference on Electrical Engineering (ICEE'14)*, Tehran, Iran (2014): 1080–1084.

48. Y. Zhang, L. Wang, and W. Sun, "Trust System Design Optimization in Smart Grid Network Infrastructure," *IEEE Transactions on Smart Grid*, vol. 4, no. 1 (March, 2013): 184–195.

49. Yichi Zhang, Weiqing Sun, and Lingfeng Wang, "Location and Communication Routing Optimization of Trust Nodes in Smart Grid Network Infrastructure," in *IEEE Power and Energy Society General Meeting*, (2012): 1–8.

50. W. Ahmad, O. Hasan, U. Pervez, and J. Qadir, "Reliability Modeling and Analysis of Communication Networks," *Journal of Network and Computer Applications*, vol. 78 (January, 2017): 191–215.

51. L. Wisniewski, "Communication reliability," in *New Methods to Engineer and Seamlessly Reconfigure Time Triggered Ethernet Based Systems During Runtime Based on the PROFINET IRT Example*, Berlin, Heidelberg: Springer, 2017, 105–123.

52. F. Al-Turjman, and A. Radwan, "Data Delivery in Wireless Multimedia Sensor Networks: Challenging & Defying in the IoT Era," *IEEE Wireless Communications Magazine*, vol. 24, no. 5, pp. 126–131, 2017.

53. P. Rohal, R. Dahiya, and P. Dahiya, "Study and Analysis of Throughput, Delay and Packet Delivery Ratio in MANET for Topology Based Routing Protocols (AODV, DSR and DSDV)," *www.ijaret.org* Issue II, vol. 1, 2013.

54. Y. Zong, Z. Zheng, and M. Huo, "Improving the Reliability of HWMP for Smart Grid Neighborhood Area Networks," in *2016 International Conference on Smart Grid and Clean Energy Technologies (ICSGCE)*, Chengdu, China (2016): 24–30.

55. Y. Qian, C. Zhang, Z. Xu, F. Shu, L. Dong, and J. Li, "A Reliable Opportunistic Routing for Smart Grid with In-Home Power Line Communication Networks," *Science China-Information Sciences*, vol. 59, no. 12 (December, 2016): 122, 305.

56. F. Al-Turjman, M. Imran, "Energy Efficiency Perspectives of Femtocells in Internet of Things: Recent Advances and Challenges," *IEEE Access Journal*, vol. 5, pp. 26808–26818, 2017.

57. A. Kannan, V. Kumar, T. Chandrasekar, and B. J. Rabi, "A Review of Power Quality Standards, Electrical Software Tools, Issues and Solutions," in *International Conference on Renewable Energy and Sustainable Energy (ICRESE'13)*, Coimbatore, India (2013): 91–97.

58. J. A. Orr and B. A. Eisenstein, "Summary of Innovations in Electrical Engineering Curricula," *IEEE Transactions on Education*, vol. 37, no. 2 (May, 1994): 131–135.

59. J. S. Subjak and J. S. McQuilkin, "Harmonics—Causes, Effects, Measurements, and Analysis: An Update," *IEEE Transactions on Industry Applications*, vol. 26, no. 6 (1990): 1034–1042.

60. M. H. J. Bollen and I. Y. Gu, "Appendix B: IEEE Standards on Power Quality," in *Signal Processing of Power Quality Disturbances*, Hoboken, NJ, USA: John Wiley & Sons, Inc., 2000: 825–827.

61. M. H. J. Bollen and I. Y. Gu, "Appendix A: IEC standards on Power Quality," in *Signal Processing of Power Quality Disturbances*, Hoboken, NJ, USA: John Wiley & Sons, Inc., 2006: 821–824.

62. F. Al-Turjman, "Optimized Hexagon-based Deployment for Large-Scale Ubiquitous Sensor Networks," *Springer's Journal of Network and Systems Management*, vol. 26, no. 2, pp. 255–283, 2018.

63. F. Al-Turjman, "Cognitive Caching for the Future Sensors in Fog Networking", *Elsevier Pervasive and Mobile Computing*, vol. 42, pp. 317–334, 2017.

64. F. Al-Turjman, H. Hassanein, and M. Ibnkahla, "Towards prolonged lifetime for deployed WSNs in outdoor environment monitoring," *Elsevier Ad Hoc Networks Journal*, vol. 24, no. A, pp. 172–185, Jan., 2015.

65. A. Al-Fagih, F. Al-Turjman, W. Alsalih and H. Hassanein, "A priced public sensing framework for heterogeneous IoT architectures," *IEEE Transactions on Emerging Topics in Computing*, vol. 1, no. 1, pp. 135–147, Oct. 2013.

66. "IEC 61000-4-30 Ed3—Power Standards Lab." [Online]. Available: https://www.powerstandards.com/testing-certification/certification-standards/iec-61000-4-30-ed3-certification-testing/. [Accessed: December, 02, 2017].

67. C. Masetti, "Revision of European Standard EN 50160 on power quality: Reasons and Solutions," in *Proceedings of 14th International Conference on Harmonics and Quality of Power—ICHQP'10*, (2010): 1–7.

68. L. Morales-Velazquez, R. de J. Romero-Troncoso, G. Herrera-Ruiz, D. Morinigo-Sotelo, and R. A. Osornio-Rios, "Smart Sensor Network for Power Quality Monitoring in Electrical nstallations," *Measurement*, vol. 103, (June, 2017): 133–142.

69. K. Koziy, Bei Gou, and J. Aslakson, "A Low-Cost Power-QQuality Meter with Series Arc-Fault Detection Capability for Smart Grid," *IEEE Transactions on Power Delivery*, vol. 28, no. 3 (July, 2013): 1584–1591.

70. P. Koponen, R. Seesvuori, and R. Bostman, "Adding Power Quality Monitoring to a Smart kWh Motor," *Power Engineering Journal*, vol. 10, no. 4 (August, 1996): 159–163.

71. K. D. McBee and M. G. Simoes, "Utilizing a Smart Grid Monitoring System to Improve Voltage Quality of Customers," *IEEE Transactions on Smart Grid*, vol. 3, no. 2 (June, 2012): 738–743.

72. F. A. S. Borges, R. A. S. Fernandes, I. N. Silva, and C. B. S. Silva, "Feature Extraction and Power Quality Disturbances Classification Using Smart Meters Signals," *IEEE Transactions on Industrial Informatics*, vol. 12, no. 2 (April, 2016): 824–833.

73. W. G. Morsi, "Electronic Reactive Energy Meters' Performance Evaluation in Environment Contaminated with Power Quality Disturbances," *Electric Power Systems Research.*, vol. 84, no. 1 (March, 2012): 201–205.

74. C. De Capua and E. Romeo, "A Smart THD Meter Performing an Original Uncertainty Evaluation Procedure," *IEEE Transactions on Instrumentation and Measurement*, vol. 56, no. 4 (August, 2007): 1257–1264.

75. M. M. Albu, M. Sanduleac, and C. Stanescu, "Syncretic Use of Smart Meters for Power Quality Monitoring in Emerging Networks," *IEEE Transactions on Smart Grid*, vol. 8, no. 1 (January, 2017): 485–492.

76. F. Al-Turjman, "Price-based Data Delivery Framework for Dynamic and Pervasive IoT," *Elsevier Pervasive and Mobile Computing Journal*, vol. 42, pp. 299–316, 2017.

77. F. G. Montoya, A. García-Cruz, M. G. Montoya, and F. Manzano-Agugliaro, "Power Quality Techniques Research Worldwide: A review," *Renewable and Sustainable Energy Reviews.*, vol. 54, (February, 2016): 846–856.

78. D. Granados-Lieberman, R. J. Romero-Troncoso, R. A. Osornio-Rios, A. Garcia-Perez, E. Cabal-Yepez, "Techniques and Methodologies for Power Quality Analysis and Disturbances Classification in Power Systems: A Review," *IET Generation, Transmission & Distribution*, vol. 5, no. 4 (2011): 519.

79. Z. Chu1, H. X. Nguyen1, T. A. Le1, M. Karamanoglu1, D. To, E. Ever, F. Al-Turjman, and A. Yazici, "Game Theory Based Secure Wireless Powered D2D Communications with Cooperative Jamming," *IEEE Wireless Days conference*, Porto, Portugal, 2017, pp. 95–98.

80. N. ul Hasan, W. Ejaz, M. Atiq, and H. Kim, "Recursive Pyramid Algorithm-Based Discrete Wavelet Transform for Reactive Power Measurement in Smart Meters," *Energies*, vol. 6, no. 9 (September, 2013): 4721–4738.

81. N. C. F. Tse, J. Y. C. Chan, W.-H. Lau, J. T. Y. Poon, and L. L. Lai, "Real-Time Power-Quality Monitoring with Hybrid Sinusoidal and Lifting Wavelet Compression Algorithm," *IEEE Transactions on Power Delivery*, vol. 27, no. 4 (October, 2012): 1718–1726.

82. G. T. Heydt, P. S. Fjeld, C. C. Liu, D. Pierce, L. Tu, and G. Hensley, "Applications of the Windowed FFT to Electric Power Quality Assessment," *IEEE Transactions on Power Delivery*, vol. 14, no. 4 (1999): 1411–1416.

83. A. Augustine et al., "Review of Signal Processing Techniques for Detection of Power Quality Events," *American Journal of Applied Sciences*, vol. 9, no. 2 (February, 2016): 364–370.

84. E. Junput, S. Chantree, M. Leelajindakrairerk, and C.-C. Chompoo-inwai, "Optimal Technique for Total Harmonic Distortion Detection and Estimation for Smart Meter," in *10th International Power & Energy Conference (IPEC'12)*, (2012): 369–373.

85. *"Measurement Practices for Reliability And Power Quality a Toolkit Of Reliability Measurement Practices,"* Ho Chi Minh City, Vietnam 2004.
86. R. Billinton and R. N. Allan, *Reliability Evaluation of Power Systems*. Boston, MA: Springer US, 1996.
87. K. Kuhi, K. Korbe, O. Koppel, and I. Palu "Calculating Power Distribution System Reliability Indexes from Smart Meter Data," in *IEEE International Energy Conference (ENERGYCON)*, Leuven, Belgium (2016): 1–5.
88. F. Al-Turjman, "Fog-based Caching in Software-Defined Information-Centric Networks," *Elsevier Computers & Electrical Engineering Journal*, vol. 69, no. 1, pp. 54–67, 2018.
89. O. Siirto, M. Hyvärinen, M. Loukkalahti, A. Hämäläinen, and M. Lehtonen, "Improving Reliability in an Urban Network," *Electric Power Systems Research*, vol. 120 (March, 2015): 47–55.
90. J. R. Hammarstron, A. da R. Abaide, M. W. Fuhrmann, and E. A. L. Vianna, "The Impact of the Installation of Smart Meters on Distribution System Reliability," in *51st International Universities Power Engineering Conference (UPEC'16)*, Coimbra, Portugal (2016): 1–5.
91. Shang-Wen Luan, J.H. Teng, Shun-Yu Chan, and Lain-Chyr Hwang, "Development of an Automatic Reliability Calculation System for Advanced Metering Infrastructure," in *8th IEEE International Conference on Industrial Informatics*, Osaka, Japan (2010): 342–347.
92. G. Singh, and F. Al-Turjman, "A Data Delivery Framework for Cognitive Information-Centric Sensor Networks in Smart Outdoor Monitoring," *Elsevier Computer Communications Journal*, vol. 74, no. 1, pp. 38–51, 2016.
93. F. Al-Turjman, "Information-centric sensor networks for cognitive IoT: an overview," *Annals of Telecommunications*, vol. 72, no. 1, pp. 3–18, 2017.

Chapter 6

Intelligent Parking Solutions in the IoT-based Smart Cities

Fadi Al-Turjman* and Arman Malekloo†

Contents

* Antalya Bilim University, Antalya, Turkey
† Middle East Technical University, Ankara, Turkey

6.1 Introduction

The idea of smart parking was introduced to solve the problem of parking space and parking management in megacities. With the increasing number of vehicles on roads and the limited number of parking spaces, vehicle congestion is inevitable. This congestion would certainly lead to a polluted environment and driver aggression, especially in peak hours where the flow density is at its maximum and locating a vacant parking spot is near to impossible. A recent report by INRIX [1] shows that on average, a typical American driver spends 17 hours a year looking for a parking space; however, looking at a major city such as New York this figure is much higher. According to the report, New York drivers spend 107 hours per year searching for parking spots and taking into account the fuel spent during this period, this results in an unnecessary emission of harmful gases which can be easily avoided.

Identifying these problems and trying to resolve them in a manner that is effective and at the same time sustainable is a challenging task, in spite of this, with the recent technological advancement, an innovative solution is proposed to once and for all solve the problems of parking spaces in cities. The Smart Parking System (SPS) is based on analyzing and processing the real-time data gathered from vehicle detection sensors such as infrared sensors, ultrasonic sensors, RFID, and many others that are placed in parking spaces to detect absence or presence of vehicles. These sensors have their benefits and weaknesses in certain areas where they are deployed. In addition, there may be issues of data anomalies where the collected information does not conform to the initial expected pattern. This could potentially lead to a less reliable system and therefore, reliability check is also another important factor that needs to be considered. Apart from these physical constraints, one should look at the security and the privacy issues of the data transmitted and received. Several factors such as end-to-end communication and data encryptions have to be considered in advance before implementing any smart applications. The data collected from the sensors may be used for several features such as, parking prediction, path optimization, and parking assignment where they could greatly help to provide an enhanced experience to both the operators of parking spaces where they could maximize their revenue and to the users of the parking to easily search, book and pay in advance.

Another approach in real-time monitoring of vehicles that has recently been discussed and gained attention is Vehicle Ad Hoc Networks (VANET), where vehicles that encompass the On Board Unit (OBU) communication system can effectively communicate with other nearby vehicles as well as roadside infrastructures to provide a real-time parking navigation service and also information dissemination of parking vacancy to and from each other [2].

A new application that is currently under study is the 5th generation telecommunication networks (5G for short) that is aiming to power the future of the Internet of Things (IoT). To put it simply, Uckelman et al. describes the IoT as the integrated part of the future Internet as a dynamic global network infrastructure with self-configuring capabilities based on standard and interoperable communication protocols [3]. As the number of connected devices increase the need for faster, better, and more efficient wireless communication is expected and this is where 5G can fill the gap [4]. The current inefficiencies in existing communication protocols in any IoT application such as SPSs are expected to be improved with the implementation of these new protocols.

The aim of this survey is to offer an insight into new researchers who seek to work in the intelligent transportation system. We look at different elements of the SPS and explain thoroughly the hardware and software aspects of this application. Therefore, we organized this survey such that Section 6.2 explains and categorizes the existing SPS by classifying them into different characteristics. Then we move onto Section 6.3 where we look at an overview of the current vehicle detection technologies where several parameters such as scalability, accuracy, and weather sensitivity are explained and compared to each other. We also introduce some of the issues in data collection and in general system reliability of SPSs in Section 6.4, by looking at different cases where the problem can arise. Section 6.5 talks about the legacy and the new long-range low power wide area network (LPWAN) communication protocols and several of their deployments in the literature. Moreover, a comprehensive table was generated where key parameters of each of these protocols are compared. Software aspects of a SPS that is responsible for managing, analyzing, and finally distributing the collected information as well as in providing enhanced features such as parking prediction and path optimization are presented in Section 6.8. Data privacy and security issues of a SPS are described in Section 6.7 and some small and large scale deployments are reviewed in Section 6.8. The VANET system which is a fairly new subject is introduced in Section 6.9 where we look at the advantages and difficulties of this new system in the smart parking application. Finally, we propose our hybrid solution in Section 6.10 where this solution can make use of many sensors with LAPWAN communication modules to provide a more reliable, fast, and secure smart parking scenario.

6.2 Smart Parking System and Classification

SPSs are categorized into various classifications in which each of them has a different purpose and use different technologies in detecting vehicles. SPSs benefit both the drivers and the operators. Drivers use the system to find the nearest parking spots and the parking operators can utilize the system and the collected information to decide on better parking space patterns and a better pricing strategy. For example, since demand for parking is not fixed, assigning dynamic parking pricing depending on the time of day and even different customers can help the operators boost their revenue [5]. Smart parking enables services such as smart payment and reservation that could essentially enhance the experience of both drivers and operators. In addition, this smart system helps to prevent vehicle theft as it increases the security measures on parking spaces, and lastly, due to minimization of the delay in finding the vacant spots which result in less number of vehicles on the road, it can greatly have a part in providing a clean and green environment by minimizing the vehicle emissions [6].

The SPS architecture layer is based on different functionalities [7, 8]: (1) the sensing layer which is the backbone of the SPS is responsible in detecting the presence or absence of a vehicle in an area using different sensor technologies, which are mostly comprised of receivers, transmitters, and anchors, (2) the networking layer is the commutation segment of the system which transmits data from transmitters or anchors to sensor gateways to the end users, (3) the middleware layer is the processing layer of any SPS which utilizes intelligent and sophisticated algorithms unique to each sensor technology to process the real-time data, and it also acts as data storage as well as a link between end users requesting services from the lower tier, (4) the application layer is the last layer in the system which is the interface of the system to the client requesting different services from different client applications, such as, mobile based application, desktop application and/or stationary information panels. The system infrastructure of the SPS is represented in Figure 6.1. In this section, the common types of SPSs are introduced, however, it should be noted that Smart Payment System (SPS), Parking Reservation System (PRS), and E-parking are considered as a complement to smart parking and they

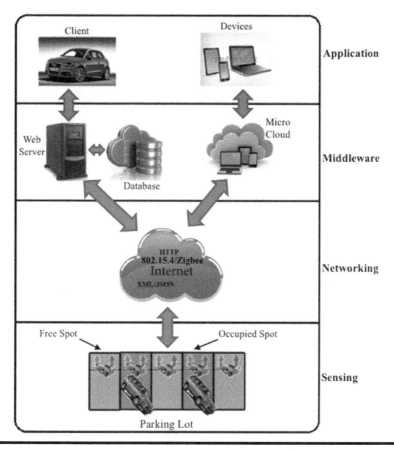

Figure 6.1   Smart parking system architecture, adapted from [8].

don't serve any function by themselves but for the sake of introduction, there are explained in this section. Lastly, Table 6.1 tabulates the parking systems based on different parameters.

6.2.1 Parking Guidance and Information System (PGIS)

PGIS, also known as Advanced Public Transport System (APTS), works by collecting parking information dynamically, mainly from loop detectors, ultrasonic, infrared, and microwave sensors to inform the drivers in real time about the vacancy of the parking space through an onboard guidance system or a Variable Message Sign (VMS) [6, 9]. PGIS consists of four major subsystems, namely: information collection, information processing, information transmission, and information distribution [10]. PGIS can be implemented both citywide or simply in parking space, where in both cases, drivers can easily follow and navigate to reach the vacant parking space [11].

Table 6.1 Different Classification of Parking Systems

Parking System Classification	Sensors	Broadcast	Guidance	Scalability	Coverage	E-parking	Reservation	Cost (O&M incl.)
PGIS	All	VMS	✓	*****	City wide, local	✓	✓	Scale dependent
TBIS	All	VMS	✓	*****	City wide, local	✓	✓	Scale dependent
CAPS	All	VMS	✓*(FCFS)	***	Local	✓	✓	*****
OAPS	Vehicle	V2V, V2I, MSN	✓	**	Local	✓	–	***
NAPS	–	–	–	*	–	–	–	None
COINS	Video & image processing	Information panel	✓	**	Local	✓	✓*	***
ABGS	All	Agent	✓	****	City wide, local	✓	✓	Scale dependent
Automated Parking	Limited	Information panel	–	–	Local	✓	✓*	****

Note: Stars (*) represent a rating system, one star (*) means *poor* and the 5 stars (*****) means *excellent* in terms of their used metric (e.g., accuracy, scalability, etc.), whereas, for the cost metric, more stars means more expensive.

The star next to the tick symbol (✓*) indicates only applicable in certain cases (applications) as discussed in each subsection.

In Hui-ling et al., a combination of PGIS and Dedicated Short-Range Communication (DSRC) is presented, where DSRC-based PGIS provides a real time, rapid, and efficient way of guidance, however, step by step implementation and concerns regarding an efficient PGIS algorithm and data safety, as well as incorporation of a different smart system may cause issues in implementing such systems [9]. Qian and Hongyan combined PGIS with a mobile phone terminal with the help of Global Positioning System (GPS) to locate and predict vacant spots and to guide drivers to the destination [10]. Whereas Shiue et al., utilized both GPS and 3G in combination [12]. Reliability of GPS and 3G connectivity in a multilevel parking space is an issue which may cause such systems to be impractical and ineffective [10].

In Chen and Chang the authors proposed PGIS in combination with ultrasonic sensors and WSN and Patil and Bhonge integrated RFID and ZigBee for smart parking solution[13, 14]. Both sensors have their disadvantages in different areas that are explained later in this chapter.

6.2.2 Transit Based Information System (TBIS)

TBIS is a park and ride based guidance system with similar functionalities to PGIS, that is, it informs drivers through VMS to guide them to vacant parking spots but it also provides real-time information of public transportation schedules and the routes which enable drivers to preplan their journey. This system not only encourages drivers to use public transportation to help reduce unnecessary vehicle emissions, it also helps to increase cities' revenue [15]. A field test in San Francisco [16] shows promising results in effectiveness of TBIS, however due to initial capital cost, such a system should be implemented mainly in large scale applications to recover system cost.

A Geographic Information System (GIS) based information system is also another way for providing information to users [17, 18]. This system provides minimum travel time by optimizing the route and schedule of the operation of public transportation in real time and enables web-based GIS systems to be implemented for the convenience of users in planning their trip.

6.2.3 Smart Payment System (SPS)

Conventional parking meters were always slow and inconvenient to use but with the advancement of technology and use of IoT, a SPS was introduced to ensure a reliable and fast method of payment [6]. This system employs contactless, contact, and mobile device technology to achieve its purpose. Such as, in contactless mode, smart cards and RFID technology such as Automated Vehicle Identification (AVI) tags, in contact mode, credit and debit cards, and lastly, in mobile devices, mobile phone services can be employed to collect payments [6, 7].

In Idris et al. the authors proposed image processing technology in conjunction with SPS by utilizing RFID technology [19]. This enables drivers to both recall their parking spot which contains the duration information of parking that also enables the calculation of parking fee. Internet-connected parking meters could also be used as a tool for parking patterns and parking predictions especially for on-street parking where the data could be valuable for predicting parking spots with the help of machine learning.

However, the technical issues with SPS are the reliability and integrity of the system in case of attacks such as signal interception or routing protocol attacks which could compromise confidential information [20].

6.2.4 Centralized Assisted Parking Search (CAPS)

In the communication of vehicles requesting parking and sensors in detecting the vacancy information, all the processing and the making of the decisions is based on a central processor (server) [21]. In this system, the First Come First Serve (FCFS) approach is adopted where the first requester is guided toward a guaranteed vacant spot (unlike other approaches such as OAPS and NAPS) closest to the driver location. However in this manner, other vehicles waiting in the queue are in constant movement until the server can satisfy them. This brings the issue of lack of cooperation of drivers that could effectively disrupt such systems. Furthermore, high initial and operating maintenance cost and scalability of CAPS are of concern [21].

Many applications of the FCFS scheduling approach is used for smart parking space management. In Kuran et al. the authors proposed a Parking Lot Recharge Scheduling (PLRS) system for electric vehicles where they compared the performance of their approach with basic scheduling mechanisms such as FCFS and Earliest Deadline First (EDF) [22]. Their optimized version of FCFS and EDF outperform the basic mechanism with regard to maximizing revenue and the number of vehicles in the parking space. In another application, edPAS [23], short for Event-Driven Parking Allocation System, focuses on effective parking space allocation based on a certain event with respect to event place with dynamic updating of the event from the communicator. This system utilized both FCFS and Priority (PR) allocation scheme where PR consistently outperforms FCFS.

6.2.5 Opportunistically Assisted Parking Search (OAPS)

Vehicles with IEEE 802.11x communication standard in ad hoc mode can share information the status and location of parking spots which enable drivers to make a more informed decision while they are searching for parking spots [21]. In this approach, drivers are guided toward the closest vacant parking space by analyzing timestamps and geographical addresses via e.g., GPS.

Since OAPS dissemination service does not impose global common knowledge of the status of parking spots, outdated timestamps and the infrequent update interval could cause delays and brings into question the effectiveness of this approach [21].

Another issue as explained in Kokolaki et al. could be the misbehavior of drivers where they enjoy the shared information from other drivers but show selfishness in sharing theirs which could increase the distance between the destination and parking spot as well as increasing parking search time [24].

6.2.5.1 Mobile Storage Node Opportunistically Assisted Parking Search (MSN-OAPS)

Instead of normal vehicle nodes, inflow of information is channeled thorough Mobile Storage Nodes (MSNs) which enables sharing of information to and from other mobile nodes acting as a relay between vehicles. Similar issues in dissemination could also be observed as the status of parking changes overtime where the accuracy of data have a tendency to drop.

In Kokolaki et al. the authors suggested that MSN, could improve the performance of OAPS however, they have no effect against facing selfish drivers [24].

6.2.6 Non-Assisted Parking Search (NAPS)

In the NAPS approach, there is no inflow of information from any vehicles or servers and the complete decision is solely depended on driver's observation in the parking space or from former experience considering the traffic flow and the time of the arrival in the parking space [21]. Drivers wander around a parking space and check for empty spots in sequence until one is found and then that spot is allocated to the driver who reached the spot first [21, 25].

6.2.7 Car Park Occupancy Information System (COINS)

COINS utilizes video sensor techniques based on one single source to detect presence or absence of vehicles. The status is then reported on information panels which are strategically placed around the parking space [26].

COINS is categorized and based on four different technologies: counter-based, wired sensor-based, wireless sensor-based, and computer vision-based, where in the latter case, it provides a more accurate result of the exact status of the parking spot without deploying any other sensors in each individual spot [26, 27].

In Bong and Lai, COINS was developed and simulated in different environments with different parameters such as weather conditions and illumination fluctuations which add an extra layer of complexity to the system [26]. Application of COINS in a multilevel parking space may not be as effective as other use of IoT technologies since scalability and coverage of such systems in these applications are the main concern.

6.2.8 Parking Reservation System (PRS)

PRS is a new concept in smart intelligent transportation systems which allows drivers to secure a parking spot particularly in peak hours prior to or during their journey [28]. The objective of a PSR is to either maximize parking revenue or minimize a parking fee. This system could easily be achieved by a simple formulation of a min–max problem.

Implementation of PRS requires several components namely: parking reservation operations information centers, a communication system between the users and the PRS, real-time monitoring of availability of parking, and an estimation of the anticipated demand [28]. Drivers can later use a variety of communication services such as SMS, mobile phone, or web-based applications to make the reservation of a parking space.

SMS based parking reservation was implemented in Hannif et al., where the integration of micro-RTU (Remote Terminal Unit) and the microcontroller and password requirement safety feature make it a smart solution in PRS [29]. Such a system is also scalable and capable of handling multiple requests from drivers. CrowdPark system is another example of PRS proposed in Yan et al., where the system works through crowdsourcing and reward points to encourage users to use the system to report parking vacancy[30]. Malicious users and accuracy of parking space are a concern in these types of systems, however, in the case of CrowdPark, a 95 percent success rate on average in both situations are reported in the San Francisco downtown area. ParkBid crowdsourced approach was proposed in Noor et al. where this system, unlike other crowdsourced applications, is based on a bidding process which provides numerous incentives about the parking spot information that enable drivers to urgently find and reserve the closest parking spot [31].

6.2.9 Agent Based Guiding System (ABGS)

ABGS or agent-based system simulates the behavior of each driver in a dynamic and complex environment in an explicit manner. An agent being an entity is capable of making decisions and has skills that make it able to define the interaction between drivers and the parking system based upon perceived facts from drivers and from different aspects such as autonomy, proactivity, reactivity, adaptability, and social ability.

SUSTAPARK was developed in Dieussaert et al. with the aim to enhance the searching experience of locating parking spaces in an urban area with an agent-based approach [32]. Their approach was to divide the problem of parking into manageable and simpler tasks that agents can follow. Another agent-based approach was introduced in Benenson et al. named, PARKAGENT based on ArcGIS with similar functionalities to SUSTAPARK but including the effects of heterogeneity of the population of drivers [33]. In Chou et al., an agent-based system was used as a negotiator to bargain over the parking fee with guidance capabilities to guide drivers to the optimal parking destination in the shortest path based on some perceived factors and interactions between other agents [34].

6.2.10 Automated Parking

Automated parking consists of computer-controlled mechanical systems which enable drivers to drive their vehicles into a designated bay, lock their vehicles, and allow the automated parking system to manage the rest of the job [6, 15]. Stacking cars next to each other with very little space in between allows this system to work in an efficient way such that the maximum available space in the parking is utilized. The retrieval process of the vehicles is easy as entering a predefined code or password. The process is fully automated which adds an extra layer of security and safety to the whole system to both drivers and vehicles which makes it the greatest benefit of such a system [15].

Although the initial cost of automated parking is very high, for the services that you receive, the price is very competitive. In fact, a 50 percent saving compared to conventional modes of parking in locations where the parking fee is high and limited is expected [27]. In these systems, employment of one or a combination of many services and sensors may be integrated to provide a fast, reliable, and a secure mode of parking with little or no interactions between the drivers and the system.

A general concern regarding such a system is that a universal building code does not yet exist and in order to serve all customers, compatibility of the system with every vehicle model is not yet possible.

6.2.11 E-Parking

E-parking as the name suggests provides a system where users can electronically obtain information about the current vacancy of parking spaces from other services and sensors, and can make reservations and payments all in one go without leaving the vehicle and prior to entering the parking space. The system can be accessed via mobile phones or the internet. In order to identify the vehicle making reservations, a confirmation code is sent to the user's email or mobile phone through SMS which can then be used to verify the identity of the vehicle [6]. The majority of the smart parking deployments that are introduced in this chapter are an example of E-parking where information about the parking vacancy can be achieved in advance. ParkingGain [35] is another example of E-parking where authors presented a smart parking approach by integrating the OBU installed on drivers' vehicles to locate and to reserve their desired parking spot. Moreover, their system offered value-added services to subsidize parts of the running cost of the system and to create business opportunities as well.

6.3 Sensors Overview

In this section, several vehicle detection technologies are discussed. They are the integrated part of information sensing in SPSs and in order to choose the best in each application several factors should be taken into account as different sensors have their strengths and weaknesses. Vehicle detection sensors are mainly categorized into two different sets [36], intrusive or pavement invasive and non-intrusive or non-pavement invasive. Table 6.2 summarizes the list of the sensors that are typically used in SPSs as well as several matrices that can influence the decision of the type of sensors to be used in smart parking applications.

6.3.1 Passive and Active Infrared Sensor

Passive infrared sensors work by detecting the temperature difference of an object and the surrounding [37], in parking applications, it identifies the vacancy of a parking space by measuring the temperature difference in the form of thermal energy emitted by the vehicle and the road. Active infrared sensors are configured to emit infrared radiation and sense the amount that is reflected from the object, therefore, detecting empty spaces in a parking space. Passive infrared sensors, unlike the other type, do not require to be anchored or tunneled into the ground or wall, but rather they are mounted to the ceiling or the ground [38]. While these sensors can accurately take advantage of multilane roads (active) and discover the exact location and speed of a vehicle on a multilevel parking space, they are prone to being affected by weather conditions such as heavy rain, dense fog and are very sensitive to the sun [39].

6.3.2 Ultrasonic

Ultrasonic sensors work in the same way as infrared sensors do but they use sound waves as opposed to light. They transmit sound energy with frequencies of between 25 to 50 kHz and upon reflecting from a vehicle, they can detect the status of parking [40] as well as several other items of useful information such as the speed of the vehicles and the number of vehicles in a given distance [41]. Like IR sensors, they are sensitive to temperature and environment, however, because of their simplicity in installation and low investment cost they are widely used in smart parking applications to identify vacant parking spaces [40, 42].

6.3.3 Inductive Loop Detector

An inductive loop detector (ILD) is normally installed on road pavements in circular or rectangular shapes [42]. They consist of several wire loops where electric current is passing through them generating an electromagnetic field with inductance which in presence or passing of a vehicle over them, they excite in frequency

Table 6.2 Comparison of Different Sensors

Sensor type	Intrusive	Count	Speed Detection	Multi-Lane Detection	Weather Sensitive	Accuracy	Cost
Passive IR	–	✓	–	–	✓	**	**
Active IR	✓	✓	✓	✓	–	**	**
Ultrasonic	–	✓	–	–	✓	***	*
Inductive Loop	✓	✓	✓	–	✓*	****	**
Magnetometer	✓	✓	✓	–	–	*****	*
Piezoelectric	✓	✓	✓	–	✓*	*****	***
Pneumatic Road Tube	✓	✓	✓	–	–	*****	*****
WIM	✓	–	–	✓	–	**	****
Microwave	–	✓	✓	✓	–	***	***
CCTV	–	✓	✓	✓	✓	****	****
RFID	–	–	–	–	–	**	*
LDR	–	✓	–	–	✓	**	*
Acoustic	–	✓	✓	✓	✓*	*	**

between 10–50 kHz resulting in a reduction in inductance that causes the electronic unit to oscillate with higher frequency which ultimately sends a pulse to the controller indicating the passing or presence of the vehicle [7, 40, 42]. Resurfacing of the road, the requirement for multiple detectors for better accuracy, sensitivity to traffic load and temperature variations, and disruption of traffic in case of maintenance are the major drawbacks of using ILD [40].

6.3.4 Magnetometer

A magnetometer functions in the same way as loop detectors. It senses the changes in earth's magnetic field that is caused by metallic objects, such as vehicles, passing over them. The cause for such distortion is that the magnetic field can flow more easily through ferrous metal than through air [43]. There are two types of magnetometers, single axis and double/triple axis magnetometers, where in the latter case the accuracy of detecting vehicles is much higher due to the fact that it uses two/three axes to identify the presence of a vehicle. Both types are unaffected by weather conditions. Lane closure, pavement cut, and in some cases short-range detection and inability to detect stopped vehicles are considered to be the shortcomings of magnetic sensors [40].

6.3.5 Piezoelectric Sensor

Piezoelectric sensors detect mechanical stress that is induced by the pressure or the vibration of objects passing over them by converting them into an electric charge. The value of the generated voltage is directly proportional to the weight of the vehicle exerting a force on the sensors. For accurate measurements multiple sensors should be used, however, they are susceptible to high amount of pressures and temperature [15, 40].

6.3.6 Pneumatic Road Tube

Pneumatic road tube sensors operate by air pressure burst which causes a switch to be closed producing an electric signal recognizing a passing vehicle. These sensors are very cost effective and offer a quick and simple installation, but in case of passing long vehicles such as a bus or large trucks over them, they lead to inaccurate axle counting resulting in less reliable parking vacancy information [40].

6.3.7 Weigh-in-Motion Sensor

Weigh-in motions (WIM) sensors can precisely determine the weight of a vehicle and the portion of the weight distributed to axles. The data gathered from these sensors are extremely useful and heavily used by highway planners, designers, as well as law enforcement. There are four distinct technologies used in WIM: load cell, piezoelectric, bending plate, and capacitance mat [40, 44]. Load cell uses a hydraulic fluid that triggers a pressure transducer to transmit the weight information. Despite its high initial investment, load calls are by far the most accurate WIM systems. Piezoelectric WIM system detects voltage variations as pressure is being applied on the sensor. Such a system consists of at least one piezoelectric sensor and two ILDs. Piezoelectric sensors are among the least expensive sensors in the market, however, their accuracy of vehicle detection is lower than that of load cell and bending plate WIM system. Bending plate uses strain gauges to record the strain or change in length when vehicles are passing over it. The static load of the vehicles is then measured by dynamic load and calibrations parameters such as speed of the vehicle and pavement characteristic. Bending plate WIM system can be used for traffic data collection. They also cost less than a load cell system, but its accuracy is not at the same level. Lastly, the capacitance mat system is consists of two or more sheet plates where they carry equal but opposite charges. When vehicles pass over them, the distance between the plates becomes shorter and the capacitance increases. The changes in the capacitance reflect the axle weight. The advantage of using this system is that it can operate for multilane roads. However, it has a high initial investment cost.

6.3.8 Microwave Radar

Microwave radar generally transmits frequencies between 1–50 GHz by the help of an antenna which can detect vehicles from the reflected frequency. Two types of microwave radar are used in this sector, Doppler Microwave Detectors and Frequency-modulated Continuous Wave (FMCW) [40, 44]. In the former type, if the source and the listener are close to each other, the listener would perceive a lower frequency whereas if they are moving apart from each other the frequency would get higher. If the source is not moving, Doppler shift would not take place. In the latter case, a continuous range of frequencies is transmitted that is changing over time. The detector would then measure the distance between the detector and the vehicle to indicate the presence of a vehicle. Microwave radars are effective in harsh weather conditions and they are also able to measure the speed of the vehicle and conduct multiple lane traffic flow data collections. However, measurement of speed in Doppler detectors requires an additional sensor to function.

6.3.9 CCTV & Image Processing

CCTV technology combined with image processing software can effectively be used in many parking spaces to determine presence or absence of a vehicle. The stream of video captured from a camera is transmitted to a computer where they become digitized and can be analyzed frame by frame using image processing software to recognize changes in the frames over time. Since CCTVs are already in place in many parking spaces for surveillance purposes, the implementation of these systems are suitable, moreover a single camera can analyze more than one parking spot simultaneously which makes it even more effective in the case of wide-area implementation; since one camera can share its information with another effectively reducing the number of cameras needed for that area [7, 42, 44]. Although CCTVs are reliable investments in SPS applications, they require to be positioned in an area where the field of view is not obstructed, and enough lighting is present. Moreover, weather conditions can affect the efficiency of these systems [28, 40, 44].

6.3.10 Vehicle License Recognition

In conjunction with CCTV and an images processing unit, vehicle plate of the ingress and egress vehicles can be captured and analyzed to give an estimation of the number of the vehicles currently present or exiting the parking spaces in real time. Furthermore, continuous monitoring of the movement of vehicles in the parking space until they reach the predesignated parking spaces that were allocated to them is a feature of these systems [28]. This also enables smart payment to be deployed which allows the drivers to exit the parking bay without any delay since the required information could be forwarded to the automated gate controller. However, bad weather conditions can disrupt the functionality of plate recognition systems. Additionally, privacy concerns regarding storing the information of vehicles on a database is another flaw of these systems.

6.3.11 RFID

RFID or Radio Frequency Identification can be used for vehicle detection in many parking spaces. RFID units consist of a transceiver, transponder, and antenna. A RFID tag or transponders unit with its unique ID can be read via a transponder antenna [42]. A RFID tag can be placed inside vehicles and the reader antenna that is placed in the parking space can read the tag and change the occupancy of the parking spot to be occupied. With this system, the delay can be minimized and flow of the traffic in the parking spaces can be smooth. Due to the range limitation of RFID, reading two tags placed on two side by side vehicles, cannot be achieved [45]. RFID in hybrid systems have shown to be effective and proved to be a more reliable option than a standalone RFID system [46].

6.3.12 LDR Sensor

A Light Dependent Resistor sensor or LDR in short, detects changes in luminous intensity. By assigning a primary source of light, such as the sun, and a secondary source such as the moon, and other surrounding light sources in the case where the sun is setting, the vehicle that is parked in the parking space creates a shadow that causes the light sensor (that is deployed in the center of the ground of the parking spot) detects luminous intensity changes, hence indicating the presence of the vehicle in the parking spot. Weather conditions such as rain, fog, and the change in the angle where the sunlight is assumed to arrive could affect the performance of these kinds of sensors and can lower the detection accuracy [47].

6.3.13 Acoustic Sensor

Acoustic sensors can detect sound energy that is produced by vehicular traffic or the interaction of tires with the road. In the detection zone of the sensor, a single processor computer algorithm can detect and signal the presence of a vehicle from the noises generated. Likewise, in the drop of the sound level, the presence of the vehicle signal is terminated. Acoustic sensors can function on rainy days and they can also operate on multiple lanes. However, cold weather and slow-moving vehicles can decrease the accuracy of such sensors. [40, 42].

6.4 Errors in Data Collection and System Reliability

Typical sensors in detecting vehicles in parking spaces may inaccurately report the number of vehicles or fail to detect at all [48]. As authors discussed in their paper, one of these errors is double counting which is apparent in the cases where drivers park their cars in between the designated parking spot causing the sensor to detect two vehicles instead of one. This behavior of drivers also causes the sensors to not report the vehicle in the spot at all. This is mainly observed in the locations where the range of the sensor is limited, therefore miscounting can also be an issue. Another type of error is occlusion where large vehicles can block the line of sight of detectors, therefore, causing the smaller vehicle next to it to be excluded. This is mostly an issue in the parking spaces where two or more vehicles are entering at the same time. The last error is phantom detection error where sensors cannot detect the correct position of the vehicles since the signals are bouncing off walls. Other than these physical constraints of detecting vehicles properly, there may be cases where the data collected from an event or observations do not conform to the initially expected patterns. Data collection is, in fact, useful when we are dealing with parking prediction and pattern analysis that gives a better management system to the operators of parking spaces and ultimately an enhanced experience to the end users. However, there is a caveat that not all the sensors can behave according to their initial expected patterns. What this brings is the issue of data anomaly or as [49] defines it; outliers. The authors defined outliers, in general, to be caused by node malfunctioning that requires inspection or maintenance. In order to identify these misconducts, they looked at a dataset provided by Worldsensing.

They discovered that by looking at the dataset as a whole, the chance of missing outliers is high. They found that there exist similar data points in the dataset that share similar characteristics and by clustering these data they were able to easily identify the outliers by applying certain sophisticated algorithms.

The majority of presented smart parking solutions do not or are unable to perform system reliability checks. The system reliability is related to, but not limited to, software, hardware, and other elements that make up the whole system that ensures an efficient and satisfactory performance of tasks in any given condition that it was designed or intended for. In Araújo et al. the authors proposed a sensor-based smart parking solution with a system reliability check [50]. They performed certain system checks to detect any errors that may be caused by hardware failure or drivers' behavior. They created false positive and false negative criteria for different time intervals to ensure the reliability of the system. In another example [51], a framework was introduced that comprised of a reputation mechanism in their mobile application based smart parking management system where each time a vacant spot is reported empty by the user, they receive a reputation score that reflects the truthfulness of the information. By doing so they ensured that the collected data is reliable and can be used as a mode of determining parking vacancy. A similar application in Villalobos et al. [52]; UW-ParkAssist, integrated the collected data from the mobile phone sensors and the official data from the parking officials or the police to reduce the data manipulation by the users. The proposed integrated system functions such that in the case of false information by the drivers, officials can manually override the status of the parking spot in the hidden settings of the application database. With this verification method, they assure that their system is able to provide improved data quality, gathered from the user's inputs.

6.5 Communication and Deployment

IoT sensors use a different number of communication protocols, however, they can all be categorized into LPWAN or short-range wireless network. LPWAN is a long-ranged and low power method of communication and the compatibility of existing cellular technology with cellular IoT (which are designed explicitly for LPWAN) makes it so that no further infrastructure is required [39]. Major LPWAN standards are, LoRaWAN, Sigfox, Weightless (SIG), Ingenu, LTE-M, and NB-IoT. Short-range WSN examples are Bluetooth, Wi-Fi, and ZigBee. A comparison of range versus bandwidth for both modern and legacy communication protocol is shown in Figure 6.2. Moreover, Table 6.3 summarizes the technical parameters for both short and long range methods of communications [53–60].

Libelium,* a wireless sensor network platform provider, has used both LoraWAN and Sigfox in their Plug & Sense platform which uses magnetic sensors to detect

---

* http://www.libelium.com/products/plug-sense/

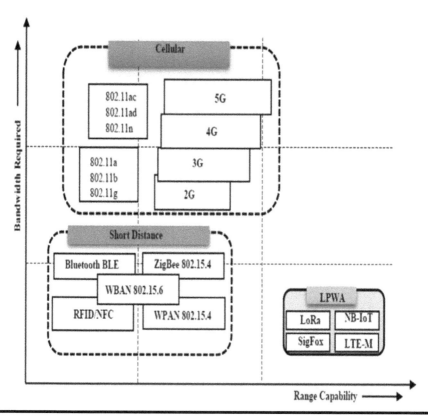

Figure 6.2   Comparison of range and bandwidth of LPWAN and other protocols [52].

vehicles in parking spots. The Huawei* solution for smart parking which resulted in 80 percent energy reduction in their Czech Republic trial and ZTE† trials in China claiming 12 percent and 43 percent reduction in congestion and time spent searching for a free spot respectively. These are some of the examples of NB-IoT based smart parking solutions. China Unicom Shanghai smart parking developed by Huawei uses a 4.5G LTE-M commination protocol in their parking network. Moreover, Nwave‡ and Telensa§ solutions to smart parking use Weightless N protocol along with magnetic sensors for detecting vehicles. A deployment of NB-IoT and third-party payment platform based SPS was studied in [61]. The authors proposed a cloud and mobile application platform that utilized a NB-IoT module that

---

* http://www.huawei.com/minisite/iot/en/smart-parking.html

† http://www.zte.com.cn/global/

‡ https://www.nwave.io/parking-technology/

§ https://www.telensa.com/smart-parking/

Table 6.3   Comparison of LPWAN and Other Technologies

Protocol	Sigfox	LoraWAN	NB-IoT	LTE-M	Weightless W/N/P	Ingenu	Wi-Fi	ZigBee	Bluetooth/ BLE
Standard/ MAC Layer	Sigfox	LORa™ Alliance	3GPP rel. 8 and 13	3GPP rel. 13	Weightless SIG	IEEE 802.15.4 k	IEEE 802.11b/g/n	IEEE 802.15.4	IEEE 802.15.1
Spectrum Bandwidth	100 Hz	125/250/500 KHz	180 KHz	1.4 to 20 MHz	5 MHz (W) 200 Hz (N) 12.5 KHz (P)	1 MHz	80 MHz (2 antennas) 20 MHz (1 antenna)	868/915 MHz	79 channels 1 MHz. 40 channels 2 MHz (BLE)
Frequency Band	ISM EU: 868 MHz US: 902 MHz	ISM 433/868/780/915 MHz	Licensed LTE bandwidth (7–900 MHz)	Licensed LTE bandwidth (7–900 MHz)	Sub GHz	ISM 2.4 GHz	2.4/5 GHz	2.4 GHz	2.4 GHz
Data Rate	100 bps UP 600 bps DL	290 bps-50 kbps	234.7 kbps UP 204.8 kbps DL	200 kbps-1 Mbps	1 kbps-10 mbps (W) 30 kbps-100 kbps (N) 200 bps-100 kbps (P)	78 kbps UP 19.5 kbps DL	11 Mbps (b) 54 Mbps (g) 1 Gbps (n/ac)	250 kbps	1 Mbps (v. 1.2) 24 Mbps (v. 4)
# Messages Per Day	140 12-byte	Defined by user	Defined by user	Unlimited	10 byte (W) Up to 20 byte (N) 10 byte (P)	Flexible (6 byte to 10 kbyte)	Unlimited	Unlimited	Unlimited
Topology	Star	Star of Star	Star	Star	Star	Star, tree	Point to hub	Star, Cluster, Tree, Mesh	Star - bus network

(Continued)

Table 6.3 (CONTINUED) Comparison of LPWAN and Other Technologies

Protocol	Sigfox	LoraWAN	NB-IoT	LTE-M	Weightless W/N/P	Ingenu	Wi-Fi	ZigBee	Bluetooth/ BLE
Battery Life	8 to 10 years	8 to 10 years	7 to 8 years	1 to 2 years	<10 years	+20 years	7 days - Up to 1 year (AA battery)	100-1000+ days	Years on coin cell battery
Power Efficiency	Very High	Very High	Very High	Medium	Very High	Very High	Medium	Very High	Very High
Range	10 km urban 50 km rural	2–5 km urban 15 km suburban 45 km rural	1.5 km urban 20–40 m rural	35 km 2G 200 km 3G 200 km 4G	5 km (W) 3 km (N) 2 km (P)	15 km urban	Up to 250 m	Up to 100 m	80–100 m
Scalability	Yes	Yes	Undetermined	Yes	Yes	Yes	Limited	Yes	Limited
Latency	1–30 s	1–2 s	1.4 to 10 s	10–15 ms	Low	>20 s	<50 ms (95th percentile)	15 ms	3 ms (BLE)
Cost	Medium	Low	High	High	Low	Medium	Low	Low	Low

can provide SMS and data transmission services over a long range with low power consumption.

As it can be observed, the applications of LPWAN protocol are limited, this is because the standard has not yet been adopted in many areas and regions. Unlike LPWAN, the legacy protocols, WSN, are the first choice in many smart parking solutions but as the population of vehicles increase, the need for longer range, more reliable, faster, and more secure mode of communication is expected. LPWAN promises to develop more to overcome the current existing challenges such as high complexity of interoperability between different LPWAN technologies, coexisting with other WSN, and lack of models for large scale applications [57, 62].

As Table 6.3 suggests, almost all the LPWAN communication modules have longer than seven to eight years of battery life with high power efficiency as well as longer range in both urban and rural areas. In rural areas, the range of the communication is longer due to less number of interferences and absence of skyscrapers, which interfere with the quality of the data sent or received. Furthermore, the most common topology used in LPWAN is of the star type where all the nodes are directly connected to a central computer or server. Every node is also connected to every other node indirectly in this topology. Short-range wireless networks are mostly connected in mesh topology to extend their range but the development cost and their energy usage for a large number of distributed devices make it ineffective in large case implementation [57], such as multilevel parking spaces. This is where LPWAN technology comes into play to overcome the limitations of the previous generations of wireless networks. Latency or the delay of the information from sensor nodes to the central server is quite small for legacy protocols compared to Sigfox or LoraWAN but this does not necessarily mean that they are not effective in SPSs. However, in large case smart parking applications where low latency is a must, NB-IoT and LTE-M are among the best options.

6.6 Software Systems in Smart Parking

Software systems plays an important role in smart parking applications. They are used to manage the collected data from the sensors, analyze, and finally distribute them efficiently. The analyzing step is based on algorithms that vary in intricacy depending on the scale and the complexity of the application. Furthermore, using the collected data and applying certain algorithms, they can be used to predict parking vacancy and also path optimization using machine learning methods that could enable the parking operators to manage their parking spaces as efficiently as possible, maximizing the parking revenue. In this section, the software aspect of SPSs and some interesting ideas such as path scheduling, path optimization, prediction, and parking assignment will be discussed.

Managing the information gathered from multiple sensors on a multilevel parking space requires a robust software system that can manage, monitor, and analyze the data in an efficient manner. Many of the examples of such a system are thoroughly discussed in Section 6.8. All of the discussed examples use a centralized server to store and manage the data. After analyzing the information, many services such as parking reservation and guidance can be implemented. In Khanna and Anand the authors presented a smart parking IoT and cloud-based system using real time information [63]. Services such as payments, reservation, and confirmation are all processed in their mobile application. In case of overshooting the parking time, their system offers extension of reservation and failing to do so will impose certain fees on the driver. A context-aware SPS based on smart server, smart object, and smart mobile was proposed in Rico et al.[64]. The smart server integrated city context information and all the related information regarding the parking status and registered users, which it then relayed to the smart object to make changes in the availability of parking space where it finally informs the user on the user interface on a smart mobile application where the user can search, book, and pay for the parking.

Guidance software systems are now becoming smart and many factors are now taken into consideration that old PGIS sacrificed. As tested in Zhu et al. the authors applied Stackelberg game theory to the PGIS so that dynamic changes of behavior of drivers can be modeled [65]. Their game strategy of balancing parking revenue of single vehicle and entire vehicles, suggests that their model can effectively reduce the average time of parking. A utility based guidance approach for parking in a shopping mall was discussed in Liang et al. [66]. Their approach made use of an improved **A*** shortest path algorithm that generates a two-stage optimal route based on user preference and the utility of parking based on six different factors. MQTT protocol parking guidance software was considered in Hantrakul et al. where the system can share real-time parking space vacancy information to at least 1,000 requests from the users [67]. Their web application in JavaScript presents the information on the layout of the shopping mall drawn by Scalable Vector Graphics (SVG). An Internet of vehicle (IOV)-based guidance system was proposed in Zhang et al. [68]. Onboard hardware in vehicles enable IOV to interact with everything around (vehicles, pedestrians, and sensors), which can then be used with optimal path and parking algorithms embedded in the system to guide the drivers to the nearest parking spot. Continuous license plate recognition using video with a Gray Level Changes algorithm and Dijkstra algorithm for locating and guiding the driver to the nearest parking spot was tested in Xie et al. [69]. The combination of license plate and Gray Level Changes detection ensure that their system is viable in case of pedestrians passing on the parking spot or when they cover the license plates. The use of GPS in smart phones with the combination of a genetic algorithm to locate and navigate to the closest parking space was developed in Aydin et al. [70]. Authors presented their solution and were able to obtain accurate results in several case studies. Instead of using exploration algorithms for locating the nearest parking space, a learning mechanism was used in Houissa et al. [71]. Authors used reinforced learning algorithms in conjunction with the Monte

Carlo approach to minimize the expected time to find a parking place in an urban area. They compared their algorithm with the method of tree evaluation and a random method and deduced that their algorithm is less complex and more efficient.

Decisions based on when and where to park are largely based on driver's observations. Many factors such as accessibility, fee, and availability of parking influence the decision of the drivers. On the other hand, decisions based on past experience when drivers locate free parking space with or without prior information of the availability always tends to lead to congestion of that particular space, increases the searching time, and causes long queues. But if the availability of the parking space can be predicted and disseminated in time, the driver's experience in locating the most suitable location can be enhanced. Moreover, prediction of parking availability offers parking operators the chance to perform short and long term system checks that ultimately enables them to take preventative decisions in case of system failure.

Nowadays, prediction is as easy as collecting data from the sensors and applying algorithms. Previously, prediction was based on historical data and surveys that were used to create a model of parking vacancy information. In Ji et al. data collected from off-street parking garages was used to create a short term model that could forecast the changing characteristic of the parking spaces by the wavelet neural network method [72]. They compared their research method with largest Lyapunov exponents in terms of accuracy, efficiency, and robustness. Although the model was successfully tested, other criteria such as driver's behavior or environmental characteristics were not taken into consideration. In another study in Caicedo et al. the authors proposed a real-time availability forecast (RAF) algorithm based on drivers' preferences and the availability of parking that are iteratively allocated using an aggregated approach [73]. Their algorithm is updated upon each arrival and departure to predict the dynamic capacity and the parking availability. Their test simulation in a parking facility in Barcelona showed promising results with small errors. The results were then compared with the numerical method where they observed no significant difference between the two approaches. Prediction of parking space availability of the Ubike system in Taipei City was tested in Leu and Zhu [74]. The authors used regression-based models (mainly linear regression and support vector regression scheme) to forecast the number of bicycles in Ubike stations. Their model, due to a constraint that the bicycles are only circulating around the station, is different from a parking system used for cars. However, a similar approach could be used for car-park facilities. A prediction mechanism based on three feature sets with three different algorithms for comparison, namely, regression tree, support vector regression, and neural networks was developed in Zheng et al. [75]. With their data sets from a San Francisco parking facility and a Melbourne parking facility and their model, they concluded that the least computational intensive algorithm, regression tree, performed better than the others.

Parking search and space optimization are another part of the software system in smart parking that utilizes the collected information to minimize the searching time and to maximize the number of parking spaces in a given parking space. The authors in Maric et al. showed that their adaptive multi-criteria optimization model effectively reduced the searching time by 70 percent in an urban area [76]. The criteria (acceptable walking distance, price, and driving time) were based on drivers' preferences that were presented by adequate utility function and the objective was to maximize the expected utility. In a study conducted in University of Akron [77], the authors used a direct random search method to perform their optimization model based on mainly different classroom assignment options. However other factors such as parking search behavior, arrival and departure distributions, and the location of different buildings and parking facilities were also taken into account. Their model managed to reduce the parking search time by about 20 percent. Cooperative parking search between vehicles searching for parking space using V2V and V2I modes of communication was studied in Rybarsch et al. [78]. The authors found that when vehicles search cooperatively, a search time reduction of up to 30 percent is observed. They also concluded that drivers would benefit more if they exchanged information both before and after reaching the destination. An intelligent hybrid model for optimizing parking space based on the Tabu metaphor and rough set-based approach was used by Banerjee and Al-Qaheri [79]. Tabu search was used as a complement to a heuristic algorithm while the rough set was used as a tool to manage noisy and incomplete data due to traffic conditions.

6.7 Data Privacy and Security in Smart Parking

It is estimated that the number of connected IoT devices will reach 20 billion by 2023 [80]. These devices are in constant communication and the large number of data that is being processed, aggregated, and shared with users can be intercepted and be used for nefarious intentions [81]. The most crucial part in any smart applications is ensuring that the network supports end-to-end encryption and authentication. Some key considerations for security and privacy protection for the integrity of the system and ultimately the users are as follows [82]: (1) data collection, which if limited to certain extent could greatly help mitigate any risks, for example, storing a large amount of data can elevate security breaches and collection of a huge amount of personal data may be used in a way that it is out of the scope of consumer's expectations from the system, (2) optimization and data analysis for any smart based application that requires information sharing, therefore, service providers and the technology partners should come to an agreement on secure data handling and techniques to ensure user's privacy such as de-identification, (3) reliability of servers, encryptions, and digital signature are also as important as risk management protocols and physical security of the system, (4) human errors, intentional or unintentional can elevate security risks, therefore, policies and procedures for training are required to mitigate oversight issues, (5) lastly, transparency of any smart system ensures the integrity of such system which offers accountably and clear policies in regards to data security and privacy.

A good smart parking application requires end-to-end communication between the end user and the server. Since the majority of smart parking solutions are established on either web-based or mobile based applications for determining the vacancy of a parking space which may also provide the ability of reservation and payment, this requires users in such systems to enter personal information such as their home/business address. Since these systems also keep track of the history of transactions including credit card information, they are therefore considered as the critical aspects in data privacy and security of the existing SPSs [81]. P-SPAN or Privacy-Preserving Smart Parking navigation system was developed by Ni et al. [83]. Their navigation system for locating and guiding drivers to a vacant parking spot uses Bloom filter and vehicular communications by introducing a privacy preservation mechanism, has been shown to be an effective SPS with low computational and communication overhead. Another VANET-based approach that is similar to the previous study with privacy-preserving in mind was discussed in Lu et al. [84]. Their system provided a secure navigation protocol with one-time credentials.

Some communication protocols lack data encryptions or require high computational resources in order to function securely; however, in Chatzigiannakis et al. the authors employed Elliptic Curve Cryptography (ECC) with LPWAN protocol as an alternative to other cryptography techniques in devices where hardware limitations exist [81]. Furthermore, in Wang and He, the authors proposed a solution for secure data collection where they used a repository of sensing information acting as a sink for the sensing data and a mirror of the reservation database which is synchronized with the repository [85]. Though this method,, drivers are the only elements that can access the mirror database for payment, for checking the vacancy of parking spaces, and reservation on their mobile devices.

6.8 Smart Parking Solutions Deployments

In this section the literature on the various implementations of smart parking based on the vehicle detection sensor as well the mode of communication and other criteria are reviewed and discussed.

A custom infrared sensor was used in Owayjan et al. to detect vehicles entering the parking spot where the results were sent to the router using Arduino by the mean wired connection, Ethernet [86]. Drivers are automatically connected to the parking network and to check for parking spaces, android mobile applications based on JSON were used to determine the vacancy. This was also used to guide drivers when they are leaving the parking spot as well as enabling smart payments. On the other hand, in Ravishankar and Theetharappan the authors used a Raspberry Pi as the central server to send the information to the cloud-based mobile application [87]. In these two examples, the scalability of these systems for multilevel parking spaces are of concern where the system reliability and effectiveness are the major points to be considered. An example of a reliable implementation is presented in Larisis where the system is comprised of both passive infrared and magnetic sensors to detect vehicles where the information was sampled in Java using TinyOS-2.x and the results were reported to drivers as web-based application [88]. A SPS for commercial stretches in cities using passive infrared sensors and image detectors was discussed in Kanteti et al. [89]. They also introduced smart payment and a reservation system in addition to guidance using GPS in their mobile application.

Ultrasonic is the most used sensor to detect vehicles in many areas. As Idris et al. presented their smart parking solution with ultrasonic sensors to detect vehicles and ZigBee standard for communication module between the sensors and the RabbitCore microcontroller which controls the multiplexers sharing the information from many sensors to the controller [90]. Their system uses shortest path algorithms to find the nearest parking spot and exit location near to the location of the driver as they enter or leave the parking space. This system requires the drivers to follow the directions given by the system and any deviation results in the failure of the system. Similarly in Kianpisheh ultrasonic was used as the method for detecting vehicles in a multilevel parking space; however, additional horizontal sensors mounted on the wall were used for detection of improper parking where an alarm would be triggered notifying the driver of his improper parking position [41]. Using three sensors per parking spot in this way would not be feasible and further studies should be conducted for cost effectiveness. Ultrasonic was also used in mobile systems. ParkNet vehicles [91] comprised of GPS receivers and mounted ultrasonic sensors on passenger side doors of vehicles which can detect parking spots in urban areas as the vehicle drives along a street. This can create a real-time map of parking information along that street.

ILDs are mostly used for traffic surveillance [39]; however, in Cheung et al. the authors tested the application for detecting vehicles and an 80 percent success rate in detecting the magnetic signature was observed [92]. IDLs are mostly used in conjunction with other sensors to increase the reliability and to provide traffic parameters such as speed, volume, and gap which could be useful in the analysis of the performance of many parking spaces.

Magnetometer sensors are also a widely used sensor in parking spaces due to their accuracy and reliability in many conditions. In Yoo et al. the authors presented their SPS that is comprised of magnetic sensors placed on each parking spot to detect the vacancy [93]. It shares the results through T-sensor to base station via a ZigBee module which later presents the parking vacancy information through VMS. Their system also offers a self-healing mechanism as well as a battery lifetime of over five years. In another example in Trigona et al. the authors used a tri-axial magnetic sensor for detecting vehicles as they enter the parking spaces [94]. Background noise produced from multiple vehicles as they approach the sensor is the issue that should be worked on. Social network car-park occupancy information was implemented in Chinrungrueng et al. where the system works by detecting vehicles via magnetic sensors placed at the entrance and exit of a parking space that can inform the drivers of the available car parks via the LED indicators placed in several locations that are connected via RS245&485 to the server and also through their Twitter account [95].

Image processing and vehicle license recognition techniques are used in combination with several smart parking implementations. In Banerjee et al., video cameras are used to detect vehicles as they enter the parking which are sequentially checked with the help of the Prewitt edge detection technique [96]. The combined use of image processing and the license recognition algorithm as previously discussed was studied in Tan et al. [97]. The system functions such that the characteristics of vehicles such as color and license plates are recorded in a database. Upon searching for parked vehicles by inputting the license plates, drivers can easily locate their vehicles. Since full recognition of a vehicle in some cases is not possible, classification probabilities based on the similarities of plates and colors is used for retrieval of vehicles. In Idris et al. the authors proposed image processing in conjunction with RFID for retrieval purposes [19]. They used a RabbitCore microcontroller and ZigBee module and they provide A* shortest path to assign a vacant spot to drivers where they can be informed of the location of the spot via VMS or the printed map on the ticket.

6.9 A New Application in the Smart Parking System

With the advancement of technology and wireless communications improvements over the years, there has been a new trend in communication technology between vehicles and infrastructures known as VANET, which is a subgroup of MANET or Mobile Ad Hoc Network that uses vehicles as mobile nodes [98]. They are classified into three different types, Vehicle to Vehicle (V2V), Vehicle to Infrastructure (V2I), and lastly, Vehicle to Roadside (V2R). The MAC layer standard of VANET is IEEE 802.11p standard with a connectivity range of 100 to 1,000 m and 27.0 Mbps data rate that promises a fast and reliable method of communication. However, they are more expensive than other protocols [99]. Onboard Units (OBU) installed in vehicles by manufacturers and the Roadside Unit (RSU) installed near roads providing both road safety and a mode of communication between the vehicles and roadside infrastructure are the requirements in these types of systems [2]. VANET is also used in SPSs where vehicles can detect parking occupancy and report the location to other vehicles or to roadside units in two-way communication [39]. As [100] presented their smart parking solution, DIG-Park, with the combination of V2I and RSU along with distance geometry algorithms to find the nearest parking spots that could be used both indoors and outdoors.. They provide real-time parking navigation as well as anti-theft and anti-collision protection that make use of OBU and RSU. Similar work was carried out by Kuran et al.; however, their model represented a large parking space and the complexity of the model was simple compared to the previous work [2].

VANET possibilities are enormous, however, some problems and difficulties arise when using this system such as [101] problems in frequency and bandwidth spectrum as VANET is operated in the band 75 MHz at 5.9 GHz where in some countries the band is used for military and radar systems. Another issue lies within the routing protocol where multi-hop communication between vehicles makes the current routing protocols not compatible with VANET, such as the current protocol in MANET, and therefore, current research is mostly focused on creating a VANET-based routing protocol.

6.10 Hybrid Solution

A hybrid model is defined as where more than one sensor is used for detecting vehicles or in the case of communications, where both wired and wireless mode are used in conjunction. Integrating multiple data source into one provides a more reliable and faster way for detecting vehicles because in the case of failure of one sensor node, the integrity of the system would not collapse. Streetline application [102] in hybrid smart parking is the perfect example of such a system. They incorporated a magnetometer, light sensors, CCTV, and a mobile phone application to detect and report the vacancy of parking spots for a large parking space. They also offer a dynamic pricing strategy for their payment systems. In another implementation of hybrid smart parking in Huang et al., the authors presented a Micro Aerial Vehicle (MAV) indoor application for a SPS which is equipped with two ultrasonic sensors and two cameras for detection of parking spots [103]. NFC tag and QR codes are also used for locating the MAV, for recording and updating of the latest status of the parking spot and as a way for drivers to record their parking information. In case of hybrid wired and wireless communication, Larisis et al. [8] presented a hybrid system that uses presence sensors and RFID tags for billing purposes which are both provided with wired and wireless capabilities, 802.15.4/ZigBee.

In general, as summarized in Table 6.4, a hybrid system provides a flexible solution to smart parking applications where the functionality of the system may change in the future. Furthermore, the scalability of such systems for a large application such as urban parking is high, and the deficiencies of a single-sensor approach in this scenario can be overcome. Hybrid systems can be the future of smart parking solutions and the worldwide usage of these systems can be anticipated in the future prospect of IoT. Although there are challenges in these systems such as they require the handling of multiple data sources entailing complex algorithms and the requirement of having a safe and fast method of communication between multiple sensors to decrease the latency of data, but in the end, the outcome of these hybrid systems is both beneficial to the user and the parking space owners as they are most effective in providing useful services such as integrated smart parking reservation, smart payments, and automatic gate control system to the drivers and dynamic

Table 6.4 Comparison of Hybrid and Single-Sensor Smart Parking System

	Parking system model	
Characteristics	*Hybrid [10, 57, 73, 75, 76]*	*Single-Sensor [3, 21, 55–65]*
Large Scale Application	+	Limited
Flexibility	+	Limited
Cost Effectiveness	+ (large scale)	Sensor Dependent
General Reliability	+	Low
Service Providers	+	+*
Overall Recommendation	+ (medium-large scale)	+ (small scale)

pricing strategies based on the availability of parking spot and other factors and dynamic parking patterns for maximum utilization of parking spaces during peak and off-peak hours to the parking space owners.

6.11 Open Issues

In this chapter, a complete survey of existing smart parking solutions was discussed where several types of sensors in terms of their functionalities and their strengths and weaknesses were introduced. As discussed in the chapter, we cannot find a single best sensor that is applicable for all the smart parking solutions. Some may fail in extreme weather or they are required to be placed inside pavements which have their own complications. Others may have limited use in certain applications due to privacy and security issues. Therefore, the efforts to diminish these shortcomings in the information collection are now put into mobile sensing devices where some of the limits or barriers of fixed sensors can be addressed. However, due to irregularities in spatiotemporal coverage and the issue of big data that are being transmitted, aggregation of such data and predicting the availability of the parking spots are the challenges, but if these challenges are tackled correctly this could be the new future of mobile sensing devices.

Moreover, having a reliable, fast, and secure mode of communication in a SPS is another problem. As discussed in this chapter, the new era of LPWAN communication protocol seems to be the future of commutations as they provide large area coverage, low power consumption, and a high battery lifetime, as well as higher security measures compared to the legacy communication modules. Large scale applications of LPWAN are still being studied as there is still the issue of interoperability and coexistence with other WSN. A large amount of data and big packet sizes that are being transmitted require steadfast, end-to-end encrypted communication. As it was estimated that by 2023, there would 20–40 billion connected devices, there could be a bottleneck in the existing communications infrastructures that affects this scenario [104].

The 5th generation mobile network is expected to be the center of the emerging IoT devices in the near future. With the ever-increasing applications in cloud computing and smart devices, 5G can promise to address, some if not all, the current issues of telecommunications. Studies are now underway to integrate existing devices with 5G wireless communication.

Connected vehicles is also another interesting option that is being used in several applications of smart parking solutions. Mobile sensors connected to the vehicles and smart mobile applications that could identify parking spots are now introduced by car manufacturers. The idea of connecting everything including our cars to each other may sound overwhelming as it can provide attractive services such as real-time navigation, crowdsourced information, and many other services. However, the current technologies in connected vehicles may limit the full potential of these types of application, particularly in urban areas [105]. Varying speed, the need for a better routing protocol, specific bandwidth for communication, and the need for high-speed communication technology to send and receive information, and in general the QoS including delays are the challenges in the current era of connected vehicles.

Nowadays, most of the deployed SPSs rely on battery-powered sensors and wireless communication modules. Although there are possibilities of implementing energy-aware algorithms or routing protocols that could effectively reduce the overall energy consumption as discussed in Lee et al.; however, at some point due to size reduction of circuit boards and increase in big data we need to look at the potential of energy harvesting–modules [106]. These modules use power generating–elements such as solar cells, piezoelectric elements, and thermoelectric elements to generate electricity by converting different energy sources such as light, vibration, and heat respectively. ParkHere [107] is a self-powered parking sensor that uses the weight of vehicles to power a Micro Generator which at the same time can send its information regarding the occupancy of the parking spot to the server via mobile radio.

6.12 Conclusion

As the population of urban area increases leading to traffic congestions and other problems, the growing need for parking spots is inevitable. Therefore, to improve the current parking system and to fix the problem of overcrowded cities lacking parking spots, smart parking was introduced. A comprehensive survey of the current SPSs including the major vehicle detection technologies used in these systems was presented and broken down according to their effectiveness, security, and communication module. The objective of this survey was to offer an insight into new research in the intelligent transportation system. We looked at different elements of a SPS and explained thoroughly the hardware and software aspect of this application. The software aspect of smart parking was presented and several features such as parking prediction, path optimization, and parking assignment and how the collected information can enhance the experience of parking operators as well as drivers was introduced. Different tables were generated which compared several key factors of each of the elements in SPS, sensors, and communication modules. Moreover, an overview of data security and privacy, as well as the new application on connected vehicles, were discussed. In the end, a hybrid solution that aimed to solve some of the current problems in smart parking application was presented. In this practice, multi-sensors hybrid smart parking can integrate multiple data from different sensors to provide a reliable, inexpensive, and efficient system. In future work, more effort should be put on researching innovative smart parking ideas. More extensive research should be conveyed on the hybrid solutions models containing state-of-the-art sensors with the integration of LPWAN and 5G communication modules, in addition to energy harvesting possibilities for a flawless SPS.

References

1. Graham Cookson, "Parking Pain—INRIX Offers a silver Bullet," *INRIX—INRIX*. Online. Available: http://inrix.com/blog/2017/07/parkingsurvey/. Accessed: November 21, 2017.
2. R. Lu, X. Lin, H. Zhu, and X. Shen, "SPARK: A New VANET-Based Smart Parking Scheme for Large Parking Lots," in *IEEE INFOCOM 2009*, (2009): 1413–1421.
3. F. Al-Turjman, and S. Alturjman, "Context-sensitive Access in Industrial Internet of Things (IIoT) Healthcare Applications," *IEEE Transactions on Industrial Informatics*, vol. 14, no. 6, pp. 2736–2744, 2018.
4. F. Al-Turjman, "5G-Enabled Devices and Smart-Spaces in Social-IoT: An Overview," *Future Gener. Comput. Syst.*, (December, 2017).
5. E. Polycarpou, L. Lambrinos, and E. Protopapadakis, "Smart Parking Solutions for Urban Areas," in *IEEE 14th International Symposium on "A World of Wireless, Mobile and Multimedia Networks" (WoWMoM'13)*, (2013): 1–6.
6. J. Chinrungrueng, U. Sunantachaikul, and S. Triamlumlerd, "Smart Parking: An Application of Optical Wireless Sensor Network," in *2007 International Symposium on Applications and the Internet Workshops*, (2007): 66. pp. 66–66.
7. G. Revathi and V. R. S. Dhulipala, "Smart Parking Systems and Sensors: A Survey," in *2012 International Conference on Computing, Communication and Applications*, (2012): 1–5.
8. A. Bagula, L. Castelli, and M. Zennaro, "On the Design of Smart Parking Networks in the Smart Cities: An Optimal Sensor Placement Model," *Sensors*, vol. 15, no. 7 (June,2015): 15, 443–15, 467.

9. Z. Hui-ling, X. Jian-min, T. Yu, H. Yu-cong, and S. Ji-feng, "The Research of Parking Guidance and Information System Based on Dedicated Short Range ommunication," in *Proceedings of the 2003 IEEE International Conference on Intelligent Transportation Systems*, vol. 2, (2003): 1183–1186.

10. Y. Qian and G. Hongyan, "Study on Parking Guidance and Information System Based on Intelligent Mobile Phone Terminal," in *8th International Conference on Intelligent Computation Technology and Automation (ICICTA'15)*, (2015): 871–874.

11. M. Buntić, E. Ivanjko, and H. Gold, ITS Supported Parking Lot Management," in *International Conference on Traffic and Transport Engineering-Belgrade*, Belgrade, Serbia, 2012.

12. Y. C. Shiue, J. Lin, and S. C. Chen, "A Study of Geographic Information System Combining with GPS and 3g for Parking Guidance and Information System," *City*, vol. 65, no. 6 (2010): 9.

13. M. Chen and T. Chang, "A Parking Guidance and Information System Based on Wireless Sensor Network," in *IEEE International Conference on Information and Automation*, (2011): 601–605.

14. Manjusha Patil and Vasant N. Bhonge, "Parking Guidance and Information System using RFID and Zigbee," *Int. J. Eng. Res. Technol.*, vol. 2, no. 4 (April, 2013).

15. M. Y. I. Idris, Y. Y. Leng, E. M. Tamil, N. M. Noor, and Z. Razak, "Car Park System: A Review of Smart Parking System and its Technology," *Inf. Technol. J.*, vol. 8, no. 2 (February, 2009): 101–113.

16. C. J. Rodier and S. A. Shaheen, "Transit-Based Smart Parking: An Evaluation of the San Francisco Bay Area Field Test," *Transp. Res. Part C Emerg. Technol.*, vol. 18, no. 2 (April, 2010): 225–233.

17. S. Pal and V. Singh, "GIS Based Transit Information System for Metropolitan Cities in India," in *The Proceedings of Geospatial World Forum*, (2011): 18–21.

18. Z. R. Peng, "A Methodology for Design of a GIS-Based Automatic Transit Traveler Information System," *Comput. Environ. Urban Syst.*, vol. 21, no. 5 (September, 1997): 359–372.

19. M. Idris, Tamil. E. M, Razak Z., Noor. N. M., and Kin. L. W., "Smart Parking System using Image Processing Techniques in Wireless Sensor Network Environment," *Inf. Technol. J.*, vol. 8, Feb. 2009.

20. N. V. Juliadotter, "Hacking Smart Parking Meters," in *International Conference on Internet of Things and Applications (IOTA'16)*, (2016): 191–196.

21. E. Kokolaki, M. Karaliopoulos, and I. Stavrakakis, "Opportunistically Assisted Parking Service Discovery: Now It Helps, Now It Does Not," *Pervasive Mob. Comput.*, vol. 8, no. 2 (April, 2012): 210–227.

22. M. Ş. Kuran A. C. Viana, L. Iannone, D. Kofman, G. Mermoud, and J. P. Vasseur, "A Smart Parking Lot Management System for Scheduling the Recharging of Electric Vehicles," *IEEE Trans. Smart Grid*, vol. 6, no. 6 (November, 2015): 2942–2953.

23. K. Raichura and N. Padhariya, "edPAS: Event-Based Dynamic Parking Allocation System in Vehicular Networks," in *IEEE 15th International Conference on Mobile Data Management*, vol. 2 (2014): 79–84.

24. E. Kokolaki, G. Kollias M. Papadaki, M. Karaliopoulos, and I. Stavrakakis "Opportunistically-Assisted Parking Search: A Story of Free Riders, Selfish Liars and Bona Fide Mules," in *10th Annual Conference on Wireless On-demand Network Systems and Services (WONS'13)*, (2013): 17–24.

25. D. Thierry, I. Sergio, L. Sylvain, and C. Nicolas, "Sharing with Caution: Managing Parking Spaces in Vehicular Networks," *Mob. Inf. Syst.*, no. 1 (2013): 69–98.
26. D. Bong, Ting K. C. and K. C. Lai, "Integrated Approach in the Design of Car Park Occupancy Information System (COINS)," *IAENG Int. J. Comput. Sci.*, vol. 35, January. 2008.
27. Mario Buntić, Edouard Ivanjko, and Hrvoje Gold, "Its Supported Parking Lot Management," presented at the International Conference on Traffic and Transport Engineering, Belgrade, 2012.
28. K. Mouskos, M. Boile, and N. A. Parker, *"Technical Solutions to Overcrowded Park and Ride Facilities,"* New Jersey Department of Transportation, FHWA-NJ-2007-011, 2007.
29. N. H. H. M. Hanif, M. H. Badiozaman, and H. Daud, "Smart Parking Reservation System using Short Message Services (SMS)," in *2010 International Conference on Intelligent and Advanced Systems*, (2010): 1–5.
30. T. Yan, B. Hoh, D. Ganesan, K. Tracton, T. Iwuchukwu, and J. S. Lee, "CrowdPark: A Crowdsourcing-based Parking Reservation System for Mobile Phones," May, 2012.
31. S. Noor, R. Hasan, and A. Arora, "ParkBid: An Incentive Based Crowdsourced Bidding Service for Parking Reservation," in *IEEE International Conference on Services Computing (SCC'17)*, (2017): 60–67.
32. K. Dieussaert, K. Aerts, S. Thérèse, S. Maerivoet, and K. Spitaels, "Sustapark: An Agent-Based Model for Simulating Parking Search," in *URISA Journal*, vol. 24, 2009.
33. I. Benenson, K. Martens, and S. Birfir, "PARKAGENT: An Agent-Based Model of Parking in the City," *Comput. Environ. Urban Syst.*, vol. 32, no. 6 (November, 2008): 431–439.
34. S. Y. Chou, S. W. Lin, and C. C. Li, "Dynamic Parking Negotiation and Guidance using an Agent-Based Platform," *Expert Syst. Appl.*, vol. 35, no. 3 (October, 2008): 805–817.
35. P. Sauras-Perez, A. Gil, and J. Taiber, "ParkinGain: Toward a Smart Parking Application with Value-Added Services Integration," in *International Conference on Connected Vehicles and Expo (ICCVE'14)*, (2014): 144–148.
36. Federal Highway Administration, "Traffic Control Systems Handbook: Chapter 6 Detectors—FHWA Office of Operations." Online. Available: https://ops.fhwa.dot.gov/publications/fhwahop06006/chapter_6.htm#t62fnb. Accessed: January 4, 2018.
37. B. Song, H. Choi, and H. S. Lee, "Surveillance Tracking System using Passive Infrared Motion Sensors in Wireless Sensor Network," in *2008 International Conference on Information Networking*, (2008): 1–5.
38. G. M. Someswar, R. B. Dayananda, S. Anupama, J. Priyadarshini, and A. A. Shariff, "Design & Development of an Autonomic Integrated Car Parking System," *Compusoft*, vol. 6, no. 3(2017): 2, 309.
39. T. Lin, H. Rivano, and F. L. Mouël, "A Survey of Smart Parking Solutions," *IEEE Trans. Intell. Transp. Syst.*, vol. 18, no. 12 (December, 2017): 3229–3253.
40. Luz Elena Y. Mimbela and Lawrence A. Klein, "A Summary of Vehicle Detection and Surveillance Technologies used in Intelligent Transportation Systems—Funded by the Federal Highway Administration's Intelligent Transportation Systems Program Office," August 2007.
41. A. Kianpisheh, N. Mustaffa, P. Limtrairut, and P. Keikhosrokiani, "Smart Parking System (SPS) Architecture using Ultrasonic Detector," *Int. J. Softw. Eng. Its Appl.*, vol. 6, no. 3 (2012): 55–58.

42. A. O. Kotb, Y. cC. Shen, and Y. Huang, "Smart Parking Guidance, Monitoring and Reservations: A review," *IEEE Intell. Transp. Syst. Mag.*, vol. 9, no. 2 (Summer 2017): 6–16.

43. M. Arab and T. Nadeem, "MagnoPark—Locating On-Street Parking Spaces Using Magnetometer-based Pedestrians' Smartphones," in *14th Annual IEEE International Conference on Sensing, Communication, and Networking (SECON'17)*, (2017): 1–9.

44. P. T. Martin, Y. Feng, and X. Wang, "*Detector Technology Evaluation*," Mountain-Plains Consortium, 2003.

45. Z. Pala and N. Inanc, "Smart Parking Applications using RFID Technology," in *2007 1st Annual RFID Eurasia*, (2017): 1–3.

46. E. Karbab, D. Djenouri, S. Boulkaboul, and A. Bagula, "Car Park Management with Networked Wireless Sensors and Active RFID," in *IEEE International Conference on Electro/Information Technology (EIT'15)*, (2015): 373–378.

47. M. Bachani, U. M. Qureshi, and F. K. Shaikh, "Performance Analysis of Proximity and Light Sensors for Smart Parking," *Procedia Comput. Sci.*, vol. 83 (January, 2016): 385–392.

48. M. Dalgleish, *Highway Traffic Monitoring and Data Quality*. Artech House, 2008.

49. N. Piovesan, L. Turi, E. Toigo, B. Martinez, and M. Rossi, "Data Analytics for Smart Parking Applications," *Sensors*, vol. 16, no. 10, September, 2016.

50. A. Araújo, R. Kalebe, G. Girão, I. Filho, K. Gonçalves, and B. Neto, "Reliability Analysis of an IoT-based Smart Parking Application for Smart Cities," in *IEEE International Conference on Big Data (Big Data)*, (2017): 4086–4091 .

51. S. Gupte and M. Younis, "Participatory-sensing-enabled Efficient Parking Management in Modern Cities," in *IEEE 40th Conference on Local Computer Networks (LCN'15)*, (2015): 241–244.

52. J. Villalobos, B. Kifle, D. Riley, and J. U. Quevedo-Torrero, "Crowdsourcing Automobile Parking Availability Sensing using Mobile Phones," *UWM Undergrad. Res. Symp.*, (2015): 1–7.

53. F. Al-Turjman, "QoS -aware Data Delivery Framework for Safety-inspired Multimedia in Integrated Vehicular-IoT," *Elsevier Computer Communications Journal*, vol. 121, pp. 33–43, 2018.

54. M. Collotta, G. Pau, T. Talty, and O. K. Tonguz, "Bluetooth 5: A Concrete Step forward towards the IoT," *ArXiv171100257 Cs*, November, 2017.

55. S. Alabady and F. Al-Turjman, "Low Complexity Parity Check Code for Futuristic Wireless Networks Applications," *IEEE Access Journal*, vol. 6, no. 1, pp. 18398–18407, 2018.

56. R. S. Sinha, Y. Wei, and S.H. Hwang, "A Survey on LPWA Technology: LoRa and NB-IoT," *ICT Express*, vol. 3, no. 1 (March, 2017): 14–21.

57. U. Raza, P. Kulkarni, and M. Sooriyabandara, "Low Power Wide Area Networks: An Overview," *IEEE Commun. Surv. Tutor.*, vol. 19, no. 2 (Secondquarter 2017): 855–873.

58. S. Al-Sarawi, M. Anbar, K. Alieyan, and M. Alzubaidi, "Internet of Things (IoT) Communication Protocols: Review," in *8th International Conference on Information Technology (ICIT'17)*, (2017): 685–690.

59. A. Asaduzzaman, K. K. Chidella and M. F. Mridha, "A Time and Energy Efficient Parking System using Zigbee Communication Protocol," in *SoutheastCon 2015*, (2015): 1–5.

60. M. Lauridsen, H. Nguyen, B. Vejlgaard, I. Z. Kovacs, P. Mogensen, and M. Sorensen "Coverage Comparison of GPRS, NB-IoT, LoRa, and SigFox in a 7800 km #x000B2; Area," in *IEEE 85th Vehicular Technology Conference (VTC Spring)*, (2017): 1–5.

61. J. Shi, L. Jin, J. Li, and Z. Fang, "A Smart Parking System Based on NB-IoT and Third-party Payment Platform," in *17th International Symposium on Communications and Information Technologies (ISCIT'17)*, (2017): 1–5.

62. A. Lavric and V. Popa, "Internet of THINGS and LoRa #x2122; Low-Power Wide-Area Networks: A survey," in *2017 International Symposium on Signals, Circuits and Systems (ISSCS)*, (2017); 1–5.

63. A. Khanna and R. Anand, "IoT Based Smart Parking System," in *International Conference on Internet of Things and Applications (IOTA'16)*, (2016): 266–270.

64. J. Rico, J. Sancho, B. Cendon, and M. Camus, "Parking Easier by Using Context Information of a Smart City: Enabling Fast Search and Management of Parking Resources," in *27th International Conference on Advanced Information Networking and Applications Workshops*, (2013): 1380–1385.

65. H. Zhu, J. Liu, L. Peng, and H. Li, "Real-Time Parking Guidance Model Based on Stackelberg Game," in *IEEE International Conference on Information and Automation (ICIA'17)*, (2017): 888–893.

66. W. Liang, Y. Zhang, J. Hu, and X. Wang, "A Personalized Route Guidance Approach for Urban Travelling and Parking to a Shopping Mall," in *4th International Conference on Transportation Information and Safety (ICTIS'17)*, (2017): 319–324.

67. K. Hantrakul, S. Sitti, and N. Tantitharanukul, "Parking Lot Guidance Software Based on MQTT Protocol," in *International Conference on Digital Arts, Media and Technology (ICDAMT'17)*, (2017): 75–78.

68. X. Zhang, L. Yu, Y. Wang, G. Xue, and Y. Xu, "Intelligent Travel and Parking Guidance System Based on Internet of Vehicle," in *IEEE 2nd Advanced Information Technology, Electronic and Automation Control Conference (IAEAC'17)*, (2017): 2626–2629.

69. L. Xie, J. Liu, C. Miao, and M. Liu, "Study of Method on Parking Guidance Based on Video," in *IEEE 11th Conference on Industrial Electronics and Applications (ICIEA'16)*, (2016): 1394–1399.

70. I. Aydin, M. Karakose and E. Karakose, "A Navigation and Reservation Based Smart Parking Platform using Genetic Optimization for Smart Cities," in *2017 5th International Istanbul Smart Grid and Cities Congress and Fair (ICSG'17)*, Istanbul, (2017): 120–124.

71. A. Houissa, D. Barth N. Faul, and T. Mautor, "A Learning Algorithm to Minimize the Expectation Time of Finding a Parking Place in Urban Area," in *IEEE Symposium on Computers and Communications (ISCC'17)*, (2017): 29–34.

72. Y. Ji, D. Tang, P. Blythe, W. Guo, and W. Wang, "Short-term Forecasting of Available Parking Space using Wavelet Neural Network Model," *IET Intell. Transp. Syst.*, vol. 9, no. 2 (2015): 202–209.

73. F. Caicedo, C. Blazquez, and P. Miranda, "Prediction of Parking Space Availability in Real Time," *Expert Syst. Appl.*, vol. 39, no. 8 (June, 2012) 7281–7290.

74. J. S. Leu and Z. Y. Zhu, "Regression-based Parking Space Availability Prediction for the Ubike System," *IET Intell. Transp. Syst.*, vol. 9, no. 3(2015): 323–332.

75. Y. Zheng, S. Rajasegarar, and C. Leckie, "Parking Availability Prediction for Sensor-enabled Car Parks in Smart Cities," in *IEEE Tenth International Conference on Intelligent Sensors, Sensor Networks and Information Processing (ISSNIP'15)*, (2015): 1–6.

76. M. Maric, D. Gracanin, N. Zogovic, N. Ruskic, and B. Ivanovic, "Parking Search Optimization in Urban Area," *Int. J. Simul. Model.*, vol. 16 (June, 2017): 195–206.

77. A. Moradkhany, P. Yi, I. Shatnawi, and K. Xu, "Minimizing Parking Search Time on Urban University Campuses through Proactive Class Assignment," *Transp. Res. Rec. J. Transp. Res. Board*, vol. 2537 (January, 2015): 158–166.

78. M. Rybarsch et al., "Cooperative Parking Search: Reducing Travel Time by Information Exchange among Searching vehicles," in *IEEE 20th International Conference on Intelligent Transportation Systems (ITSC'17)*, (2017): 1–6.

79. S. Banerjee and H. Al-Qaheri, "An Intelligent Hybrid scheme for Optimizing Parking Space: A Tabu Metaphor and Rough Set based Approach," *Egypt. Inform. J.*, vol. 12, no. 1 (March, 2011): 9–17.

80. "Internet of Things Outlook—Ericsson," *Ericsson.com*, 09-Nov-2017. Online. Available: https://www.ericsson.com/en/mobility-report/reports/november-2017/internet-of-things-outlook. Accessed: December 26, 2017.

81. F. Al-Turjman, and S. Alturjman, "Confidential Smart-Sensing Framework in the IoT Era," *The Springer Journal of Supercomputing*, vol. 74, no. 10, pp. 5187–5198, 2018.

82. F. Al-Turjman, and S. Alturjman, "Context-sensitive Access in Industrial Internet of Things (IIoT) Healthcare Applications," *IEEE Transactions on Industrial Informatics*, vol. 14, no. 6, pp. 2736–2744, 2018.

83. J. Ni, K. Zhang, Y. Yu, X. Lin, and X. S. Shen, "Privacy-preserving Smart Parking Navigation Supporting Efficient Driving Guidance Retrieval," *IEEE Trans. Veh. Technol.*, vol. PP, no. 99 (2018): 1–1.

84. R. Lu, X. Lin, H. Zhu, and X. Shen, "An Intelligent Secure and Privacy-preserving Parking Scheme through Vehicular Communications," *IEEE Trans. Veh. Technol.*, vol. 59, no. 6 (July, 2010): 2772–2785.

85. H. Wang and W. He, "A Reservation-based Smart Parking System," in *IEEE Conference on Computer Communications Workshops (INFOCOM WKSHPS)*, (2011): 690–695.

86. M. Owayjan, B. Sleem, E. Saad, and A. Maroun, "Parking Management System using Mobile Application," in *Sensors Networks Smart and Emerging Technologies (SENSET)*, 2017, 1–4.

87. F. Al-Turjman, "Fog-based Caching in Software-Defined Information-Centric Networks," *Elsevier Computers & Electrical Engineering Journal*, vol. 69, no. 1, pp. 54–67, 2018.

88. N. Larisis, L. Perlepes, G. Stamoulis, and P. Kikiras, "Intelligent Parking Management System Based on Wireless Sensor Network Technology," *Sens. Transducers*, vol. 18 (January, 2013): 100–112.

89. D. Kanteti, D. V. S. Srikar, and T. K. Ramesh, "Smart Parking System for Commercial Stretch in Cities," in *International Conference on Communication and Signal Processing (ICCSP'17)* (2017): 1285–1289.

90. M. Y. I. Idris, E. M. Tamil, N. M. Noor, Z. Razak, and K. W. Fong, "Parking Guidance System Utilizing Wireless Sensor Network and Ultrasonic Sensor," *Inf. Technol. J.*, vol. 8, no. 2 (February, 2009): 138–146.

91. S. Mathur et al., "ParkNet: Drive-by Sensing of Road-side Parking Statistics," in *Proceedings of the 8th International Conference on Mobile Systems, Applications, and Services*, New York, NY, USA, 2010, 123–136.

92. S. Y. Cheung, S. C. Ergen, and P. Varaiya, "Traffic Surveillance with Wireless Magnetic Sensors," in *Proceedings of the 12th ITS World Congress*, vol. 1917 (2005): 173–181.

93. S. Yoo et al., "PGS: Parking Guidance System Based on Wireless Sensor Network," in *3rd International Symposium on Wireless Pervasive Computing*, (2008): 218–222.

94. C. Trigona et al., "Implementation and Characterization of a Smart Parking System Based on 3-axis Magnetic Sensors," in *IEEE International Instrumentation and Measurement Technology Conference Proceedings*, 2016, 1–6.

95. J. Chinrungrueng, S. Dumnin, and R. Pongthornseri, "iParking: A parking Management Framework," in *11th International Conference on ITS Telecommunications*, (2011): 63–68.

96. S. Banerjee, P. Choudekar, and M. K. Muju, "Real Time Car Parking System using Image Processing," in *3rd International Conference on Electronics Computer Technology*, vol. 2 (2011): 99–103.

97. H. C. Tan, J. Zhang, X. C. Ye, H.Z. Li, P. Zhu, and Q. H. Zhao, "Intelligent Car-Searching System for Large Park," in *International Conference on Machine Learning and Cybernetics*, 2009, vol. 6, 3134–3138.

98. H. Y. Chang, H. W. Lin, Z. H. Hong, and T. L. Lin, "A Novel Algorithm for Searching Parking Space in Vehicle ad hoc Networks," in *Tenth International Conference on Intelligent Information Hiding and Multimedia Signal Processing*, (2014): 686–689.

99. Ms.M. Santhiya, M. M. S. Karthick, and Ms. M. keerthika, "Performance of Various TCP in Vehicular ad hoc Network Based on Timer Management," *Int. J. Adv. Res. Electr. Electron. Instrum. Energy*, vol. 2, no. 12 (December, 2013): 6160–6166.

100. W. Balzano and F. Vitale, "DiG-Park: A Smart Parking Availability Searching Method using V2V/V2I and DGP-class Problem," in *31st International Conference on Advanced Information Networking and Applications Workshops (WAINA'17)*, (2017): 698–703.

101. C. Jeremiah and A. J. Nneka, "Issues and Possibilities in Vehicular Ad-hoc Networks (VANETs)," in *2015 International Conference on Computing, Control, Networking, Electronics and Embedded Systems Engineering (ICCNEEE'15)*, (2015): 254–259.

102. "Streetline," *Streetline*. Online. Available: https://www.streetline.com/. Accessed: December 26, 2017.

103. C. H. Huang, H. S. Hsu, H. R. Wang, T. Y. Yang, and C. M. Huang, "Design and Management of an Intelligent Parking lotT System by Multiple Camera Platforms," in *IEEE 12th International Conference on Networking, Sensing and Control*, (2015): 354–359.

104. F. Al-Turjman, "Modelling Green Femtocells in Smart-grids," *Springer Mobile Networks and Applications*, vol. 23, no. 4, pp. 940–955, 2018.

105. F. Al-Turjman, "Mobile Couriers' Selection for the Smart-grid in Smart cities' Pervasive Sensing," *Elsevier Future Generation Computer Systems*, vol. 82, no. 1, pp. 327–341, 2018.

106. P. Lee, H. P. Tan, and M. Han, "Demo: A Solar-powered Wireless Parking Guidance System for Outdoor Car Parks," in *Proceedings of the 9th ACM Conference on Embedded Networked Sensor Systems*, New York, NY, USA, (2011): 423–424.

107. F. Al-Turjman, "Modelling Green Femtocells in Smart-grids," *Springer Mobile Networks and Applications*, vol. 23, no. 4, pp. 940–955, 2018.

Chapter 7

Intelligent Medium Access for Adaptive Speed Limits in the Smart-Cities' Framework

Fadi Al-Turjman*

Contents

* Antalya Bilim University, Antalya, Turkey

129

7.1 Introduction

Every year, several people die in car accidents/collisions. For example, over 1.2 million died in traffic accidents around the world in 2016 [1]. There are a number of different reasons for road accidents/collisions in different countries. Nevertheless, one of the major causes of these road collisions is driving the car under unpredictable weather conditions. It was noted that fifty percent of the fatal collisions happen while driving under a speed of 55 km/h [1]. We therefore need a system where the speed limits are set according to the current weather, traffic, and road conditions. With the recent revolution in wireless telecommunications, several advanced solutions relying on wireless communication standards have been proposed to provide Intelligent Transportation Systems (ITS) in the Internet of Things (IoT) paradigm. For instance, Sahoo et al. projected an Automatic Speed Control (ASC) system that adjusts the speed of the vehicle according to the speed limit on the road [2]. The feasibility of a smart box called "Telematics," which has the ability to capture, analyze, and communicate, is being studied in cooperation with IBM's Engineering and Technology Services. Using multiple microprocessors and tiny sensors attached to the vehicle body, it is able to observe the vehicle's velocity for example and compare it to the upper speed limit of the road. In the case that the speed of the car is higher than the announced limit allowed by authorities, the box will verbally notify the driver. Moreover, a digital image processing system has been proposed by Baró et al. [3] that utilizes onboard cameras to read and recognize signs at the side of the road and send the warning signal to the driver and/or directly control the car. Different versions of this system have been investigated intensively all over the world. Results in van de Beek et al. [4] have shown that this solution is able to cut down the rate of accidents by 35 percent. In the near future, the speed control system will be very dependent on the standard of IEEE 802.16 to locate each vehicle and satisfy the demands of the required real-time services such as voice and video. The standard of IEEE 802.16 is designed and developed to offer specific services for wireless radio interface [5]. The Task Group (TG) of IEEE 802.16 has significant advantages, namely, higher data rate, scalability, real-time serviceability, low-cost maintenance, and cost upgrade [6]. However, the IEEE 802.16 d/e does not have an appropriate scheduler algorithm for the real-time service [5]. This research focuses its attention on the IEEE 802.16 system to enhance the reliability of the emerging vehicular networks. Since an IEEE 802.16 system cannot be distinctly specific in its resource allocation among the real-time applications, the base station unfairly allocates and shares the resources, namely, frequency spectrum and time slots with different types of traffic flows [5].

In IEEE 802.16, there are different Medium Access Control (MAC) scheduling services, such as Unsolicited Grant Service (UGS), real-time Polling Service (rtPS), and non-real-time Polling Service (nrtPS) to provide better quality of services. UGS and rtPS are the two schedulers for the real-time traffic. Each real-time scheduling mechanism has a parameter to quantify its bandwidth requirements, namely, delay, and the minimum and maximum transmission rates [7]. However, these schedulers do not fulfill the requirements for the real-time services in smart-cities. Hence, the suitability of a Batch Markovian Arrival Process (BMAP) is analyzed in studies such as Klemm et al. [8] for modeling of IP traffic, and it is shown that the BMAP model is a better candidate especially compared to other popular processes such as the Markov Modulated Poisson Process. Thus, we propose a Real Time-BMAP (RT-BMAP) model for real-time services. The objective of RT-BMAP is to achieve the required quality of service with the minimum delay. Accordingly, the major contributions of this chapter are as follows:

1. We present the Enhanced-real time Polling System (E-rtPS) with RT-BMAP and proactive resource allocation framework to resolve the real-time network traffic issues and solve the interference problem optimistically.
2. The proposed proactive resource allocation framework offers less computational overhead and yet again well-nigh performance in terms of resource allocation to maximize the transmission rate of each user.
3. The examination of real-time results shows that the proposed framework outperforms the existing IEEE 802.16 services in terms of throughput, session setup delay, and packet reception ratio.

7.2 Related Work

In the LTE-A, QoS Class Identifier (QCI) uses Guaranteed Bit Rate (GBR) and non-Guaranteed Bit Rate (non-GBR) to support different priorities and delay requirements [9]. According to the statistical report generated for mobile data traffic in 2013, the global usage of mobile data has grown dramatically to 83 percent [8]. The datum report says that the data traffic reached 1.5 exabytes at the end of 2013, which was a hike of 820 petabytes per month at the end of 2012. As reported in CISCO forum, the network applications on global networks grew the data traffic 63 percent in 2016, which was 7.2 exabytes per month in 2016 and 4.4 exabytes per month in 2015. In fourth generation (4G) networks, mobile data traffic has exponentially surged to 69 percent, which is four times more data traffic than third generation (3G) connections. Since smart phone usage has risen to half a billion phones, mobile devices and connections account for 89 percent of the usage of mobile data traffic. So, in 2016, smart phones generate 13 times more data traffic than non-smart phones. The CISCO statistical survey reports that the average traffic rate per smart phone was 1,614 MB per month in 2016, whereas it was about 1,169 MB per month in 2015.

The usage of mobile video service is predicted to increase ninefold in the interval between 2016 and 2021. The review of CISCO 2016 reports that the mobile data traffic is growing exponentially in the Middle East and Africa (about 96 percent), followed by Asia Pacific (about 71 percent), Latin America (about 66 percent), Central and Eastern Europe (about 64 percent), Western Europe (about 52 percent), and North America (about 44 percent) in 2016.

Since the system resources are limited and sometimes this causes a violation of QoS, the base station should proactively allocate the available bandwidth to the networking systems. IEEE 802.16 and LTE-A standard has not had any native scheduling algorithms for the effective usage of bandwidth. Therefore, it is an open challenge for individual venders to implement the scheduling mechanism [10].

7.3 Framework Description

In this section, we describe a complete vision of the proposed Vehicular-Cloud and IoT framework. Framework components are defined in the following list:

- Management & Control Venter (MCC): This component controls the velocity of the vehicle depending on the street conditions, traffic, and surrounding environment conditions.
- Speed Limits Transmitters (SLT-x): This component is used to communicate the speed limit to the end-driver. The speed limits are transmitted as a wireless message. The MCC controls the speed of the vehicle, and adapts it depending on the street, traffic, and weather conditions, given that enough Road Side Units (RSUs) are located at predefined points on the road.
- Speed Limits Receiver (SLR-x): This component takes the transmitted street upper limit and displays it clearly in the vehicle. It can also be communicated on the driver's smartphone.
- Vehicle Speed Sensor: This component is required to measure the speed of the vehicle accurately. The speed measurement system that already exists in the car can be used as well.
- In-vehicle Micro Controller: The main task of this component is to compare the actual speed to the road speed limit, which is received by the SLR-x. Based on the vehicle speed, the controller may generate an audio warning or may communicate such an incidence to the MCC through the 4G/5G network via the driver's smartphone.
- 4G/5G Modem: This component is used to send speed data to a central station which is monitored and operated by any other governmental or private entity.
- Drivers' Records Server (DRS): This component stores information about the drive. The DRS is updated when the GSM modem sends a signal from the driver's car. The DRS can be accessible by third parties with the approval of the driver. Such third parties include parents, family members, and insurance companies.

Table 7.1   Simulation Parameters of IMS Networks

Parameters	Values
Frequency	2 GHz
Bandwidth	5 MHz
Number of Cells	19
Distance (Inter-Site)	500 m
Number of RBs	25
Shadowing Standard Deviation	8 dB
Uplink Device Transmission Power	24 bBm
Proximity Distance	10 m
Maximum Power Transmission P	24 dBm
Threshold η	0.8
Channel odel	200Tap, Urban [9]
Path loss [Distance-Dependent]	$128.1 + 3.76 \log$ (R), R in km
Modulation and Coding Scheme	QPSK, 16 QAM, 64 QAM (1/2, 2/3, 3/4)

7.3.1 Functioning of the System

Figure 7.1 shows a simplified schematic for the planned Vehicular-cloud system, in normal conditions. The SLT-xs are connected to the MCC component through the Public Switched Telephone Network (PSTN) and the GSM/4G/5G networks, which are usually connected to Cloud data centers. The SLT-xs are controlled by the MCC, which recommends changing the speed limit depending on the road, weather, and traffic conditions. The MCC gets the deciding information from the patrol police, forecast stations, and driver's smart phones. The vehicle speed is continuously compared to the received speed limit which is also displayed to the driver. Hence, the driver will always know the upper limit of the street they are on. This is a better method than the conventional speed limit signs located on the side of the road. If a driver is driving below the speed limit, there is nothing to do because it might be a case of congestion. If the vehicle speed limit exceeds the received one, a warning signal will be generated to alert the driver that they exceeded the speed limit. If the driver does not respond within a given period, the In-vehicle Micro Controller sends a note (violation) through the 4G/5G modem to the DRS.

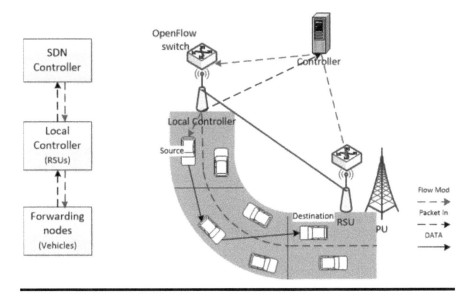

Figure 7.1   A schematic for the vehicular-cloud in smart-cities.

The violation can also be sent to the driver's smart phone. The violation can also be recorded using various network-positioning techniques. Other forms of violations such as tail tracking and red-light tracking can be motored in the car which has other types of sensors. In such a case where a violation was recorded, the DRS can inform the driver instantaneously through their mobile/e-mail/mail, or by some other IEEE 802.16 standard means. We remark that this study focused on the Voice over IP (VoIP) service as an example of the exchanged real-time contents over the vehicular-cloud.

Existing service categories supported by 802.16, UGS, and rtPS are designed and developed for the support of real-time (e.g., VoIP) communication flow. The existing and proposed real-time services can be concisely described as follows:

Service of UGS: This service is intended to provide static-size packet flow to the real-time applications, such as E-Carrier(E1)/T-Carrier(T1)/Integrated Services Digital Network (ISDN) technologies, UCTIMS client and real-time services such as VoIP (without silence detection). To send the voice data packet, the Base Station (BS) usually communicates the static size to the Mobile Subscriber Station (MSS). When the real-time application integrates the voice codec with silence detector, the bandwidth consumption of the application should be less for the off period. Otherwise, it wastes the resource availability.

Service of rtPS: The service of rtPS is intended to provide variable-size data packet flow to real-time applications like MPEG-Video Streaming periodically [5]. The BS puts up a request of bandwidth as polling and the MSS sends a report of bandwidth demand through the request of BS, thus, the process of bandwidth request incurs an additional MAC overhead and queuing-delay. The voice connection process can be negotiated in the requisition process; and called in a polling or bandwidth request process. As the rtPS service always relies on the bandwidth request process as a suitable grant size, it is able to transport the voice data more efficiently than the UGS service. However, this service causes connection delay and MAC overhead. In addition, the MSS uses a piggyback request to grant VoIP services, since the voice service is delay sensitive. In this study, we propose an improved real-time service support E-rtPS.

7.3.2 Registration Phase

RSUs are located in different places along a given road. Local Controllers (LC) are static base stations in the city. Mobile Controllers (MC) are sensing platforms attached to vehicles. We use a Minimum Spanning Tree (MST) algorithm to select an LC at each route. Each MC is considered as a node in the MST, while all RSUs are considered as terminal nodes, as depicted in Figure 7.2. Each leaf node is in different layers from the MC. RSUs in the same depth communicate with each other and select one as LC. This registration phase is summarized in the following four main steps:

Step 1: The MC first transmits a control message such as a Hello message to all RSUs which are within communication range to determine the layout of the network. As shown in Figure 7.1, if only two nodes are within the range of the MC, then these nodes are the first to receive the message.

Step 2: In reception of a message by each RSU, it computes its delay and compares the delay with the neighboring RSUs in the same depth. The RSU with the lowest delay declares itself as the LC.

RSUs can calculate the number of Hello messages sent previously as the vehicle moves and updates each other and the nearby RSUs. This helps improve the stability of the system by averting sparse network conditions, and thus, selecting a LC with the highest connectivity. We calculate the HopCount by looking at the number of layers and RSUs a message has gone through to get to the MC. (see Figure 7.2.)

Step 3: The selected LC broadcasts a message to the MC and all RSUs in range, with the updated information about all the nodes; these are vehicles and RSUs. All nodes contacted record the route to an MC in their flow table.

Step 4: The previous three steps are repeated at every step, and finally, the MC institutes a global layout and sends it to all LCs, RSUs, and vehicle networks.

At the end of this phase, the selected LCs create a localized global view at each depth from the MC. Therefore, different kinds of controllers are used to reduce the system burden form the single main controller and reduce overall overhead and delay.

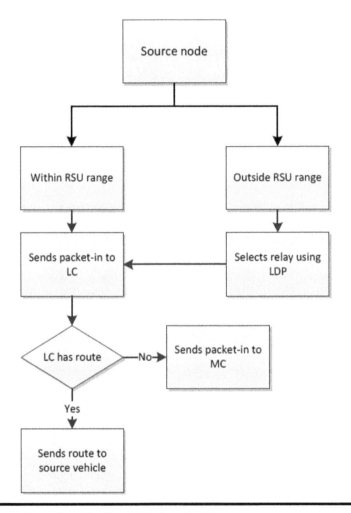

Figure 7.2   Flow chart for the vehicular-cloud.

7.4 Enriched-rtPS (E-rtPS) with RT-BMAP

Since the services of UGS and rtPS do not support the method of on/off to infer whether the state of Voice Activity Detector (VAD)/Silence Detector (SD) is active or not, the VoIP service is not considered to be efficient. In addition, the service of UGS incurs the resource wastage during off period whereas the service of rtPS incurs the MAC overhead and queuing-delay. To resolve these issues, the E-rtPS employs a VAD/SD for the voice codec. To perform well, the IP real-time subsystem-network (IRS-N) uses channel quality information channel (CQICH) to demand bandwidth from available wireless channel resources. The generic MAC header of IEEE 802.16 has two reserved bits to do additive operation, and one of them is used to inform the BS about the state of MSS voice transition. This reserved bit is called 'Grant Size (GS)' bit. When the voice connection of the MSS is 'on,' the MSS assigns the GS bit to '1'; otherwise it assigns its GS bit to '0'. Most importantly, the MSS informs the BS about its change in voice transition state without MAC overhead, since it imparts its change of transition using a traditional generic MAC header. The generic MAC header only transmits the voice transition state between the MSS and BS; however, the bandwidth request process related to IEEE 802.16 system header is not transmitted. Instead, the bandwidth request process can be gained through the uplink resources, while the MSS and BS are monitoring the GS bit transmission.

Real Time-Batch Markovian Arrival Process (RT-BMAP): The performance degradations (packet delay, MAC overhead, and resource wastage) for UGS and rtPS services, are mainly caused because the BS assignment regards the uplink resource with the consideration of voice transition to the IRS-Network (IRS-N). To infer the state of the voice/real-time transition, in this study, we integrate the technical feature of VAD/SD with the IRS client codec. While the feature of VAD/SD is integrated with the IRS client, the IRS-N can deduce whether the voice state transition is on/off. The deduction of such a transition is usually performed in the higher-layer and the layer of MAC is used to find the primitives of convergence for sub-layers from the standard system of IEEE 802.16d/e. Since the IRS-N informs the BS regarding the state of the voice transition, the IRS-N necessitates the method of on/off to infer the current status of voice transition. We thus decided to utilize the reserved bits of IEEE 802.16d/e to deduce the status of voice transition. To probe the reserved bit realistically, we represent the bit as Status-Check (SC). While the voice state transition is going 'on,' the IRS-N represents the SC bit as '1'; otherwise, the IRS-N sets the SC bit as '0'. The significant use of SC is to avoid the MAC overhead. The IRS-N employs the bandwidth request to hold the usage of uplink resource and the frame of voice codec incurred in the IRS-N determines it. Since the incursion of codec frame depends on the IRS-N, the generation of voice packets relies on the communication duration of IRS clients. The reserved bit of SC controls the voice state transition of the IRS clients through the knowledge of IRS-N to prevent occurrence of MAC overhead. In this study, in order to analyze the SC operation, a real-time client server system is employed. The IEEE 802.16 system model is constructed in the MAC layer to utilize UGS, rtPS, and E-rtPS as the on-demand resource for VoIP services similar to the studies in [10, 11, 12]. Real-time Client (RC) and Real-time Server (RS) are deployed as the real-time agents to support the voice call establishment and termination. The real-time agents use SIP message transmission using Real-time Transport Protocol (RTP)/User Datagram Protocol (UDP).

7.5 Results and Discussions

In this section, we present the results of our planned framework. It involves hardware and software implementation parts. The main purpose of the hardware part is to construct a simple testbed for such projects/ideas. This testbed has been proved to be very helpful with the cost approximation and it provides critical information about design challenges such as the delay and system throughput. This simple test-bed is composed of three subsystems, the in-vehicle subsystem connected via cellular networks, the IP-based network (Internet), and the Management and Control Center (MCC) represented by the IRS-N client. We remark here that we assumed there were 10 cars while obtaining the results in this section for realistic verification purposes. The MCC is implemented via Ubuntu-PC (laptop) for experimental purposes. And the car-toys were equipped with Arduino boards attached to GPRS modules. To evaluate the performance of the assumed resource allocation framework in the central cell, we consider using 5 MHz $\langle lb \rangle$ LTE-A with 19-cells working at 2.0 GHz as a wireless cellular network. The bandwidth of 5 MHz LTE-A is split into 25 RBs [9]. As the cell has 500 meters as an inter-site distance, the network users can be randomly distributed within the cell site [9]. For each user, the uplink power transmission is set to be 24 dBm and the threshold value η of interference avoiding mechanism is assumed to be 0.8 [9]. The selection mode of Modulation and Coding Scheme (MCS) depends on the RBs allocation for the users with minimum Signal to Noise Ratio (SINR). The mapping lookup table for SINR value to MCS mode is found in [9]. For each user, the packet arrival process follows the poison distribution with the mean arrival rate 1000 bits/sec. Table 7.1 summarizes the simulation parameter of IMS networks [9].

In order to examine the quality of online (real-time) services, namely, UGS, rtPS and E-rtPS with RT-BMAP, an IRS core function with Proxy Call Session Control Function (P-CSCF), Interrogating Call Session Control Function (I-CSCF), Serving Call Session Control Function (S-CSCF), and Home Subscriber Station (HSS) has been installed in Ubuntu supporting LTE, as shown in Figure 7.3. This testbed also integrates TS 23.167 [13], which has real-time service architecture not only to allocate the network resources but also to ignore the roaming restriction as referred to in [14]. Since this architecture is planned for ITS systems where high safety levels are required, we assume the registered requests as Emergency-Calling (E-Calling) that contains an initial request to grant the emergency services as "Unrecognized E-Calling [15]".

To probe the voice connection as a real-time service, the packet generator is integrated with the IRS core function to generate the concurrent voice calls. These voice calls can be transmitted in either serial or parallel to analyze the effectiveness of E-rtPS with RT-BMAP in comparison with UGS and rtPS. Since the IRS core function can extend its service scalability for the device connection through the knowledge of 3GPP networks [9], we have deployed an IRS core network, which has a feature of LTE-Advanced to enlarge the transmission region and provide better service flexibility. In this study, the IRS core is embedded with its Call Session Control Functions (CSCF) in order to have a realistic testbed. Besides, this core network integrates the HSS to cross-examine the stability of voice service over available networks. As the service network is defined for the real-time connectivity, we increase the voice connectivity gradually for the analysis of connectivity delay

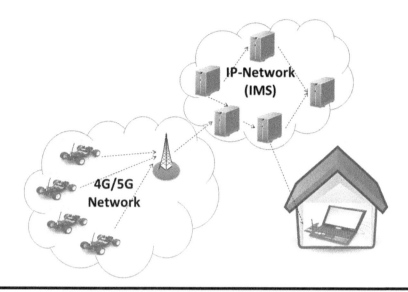

Figure 7.3   Experimental testbed for the vehicular-cloud system of IEEE 802.11d/e.

and service throughput rate. An emergency voice call session is defined as the generation of anonymous voice calls to examine the session setup delay and throughput rate of the real-time systems.

For the establishment of VoIP call and service connection, we employ the IRS client (known as UCTIMS) and packet analyzer (known as Wireshark). This experimental setup aims not only to evaluate the SIP based VoIP performance over IRS core using wireless connectivity but also to examine the voice connectivity delay and service throughput rate of the networks.

As illustrated in Figure 7.4, the throughput of the proposed algorithm is significantly higher in comparison to UGS and rtPS. In addition, as shown in Figure 7.5, the packet delay of the proposed algorithm is also superior for all the critical regions when compared with the other services. This behavior is evident even when the packet delay is set as 60 ms. Please note that the predetermined delay value is an important criterion in the system of IEEE 802.16 d/e [5] for the packets with delay violation.

7.6 Concluding Remarks

This chapter presents an adaptive framework for dynamic speed management in smart cities. The system makes use of the latest development in wireless communication and exploits the existing telecommunication infrastructures which have been used in data streaming, sound, and video to maximize the system adaptability and reduce the price. A key component to the planned framework is the dynamic

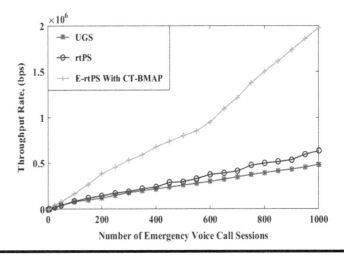

Figure 7.4   Throughput versus number of voice call sessions.

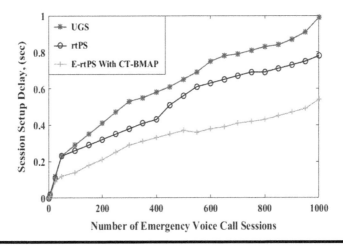

Figure 7.5   Average packet delay versus number of voice connections.

medium access approach for real-time communications. Accordingly, this chapter proposes the E-rtPS for the vehicular-cloud of IEEE 802.11 d/e. The proposed approach integrates RT-BMAP to analyze the throughput rate and average packet delay. For practical assessment, the real-time system is configured with the IEEE 802.16 d/e services, namely UGS, rtPS and E-rtPS with RT-BMAP. The experimental results clearly show that the proposed approach is more efficient in terms of average packet delay, as well as throughput when compared with the results obtained from existing 802.16 d/e services in the literature.

References

1. Research and Innovative Technology Administration, Bureau of Transportation Statistics, National Transportation Statistics, www.bts.gov/publications/national_transportation_statistics.
2. F. Al-Turjman, "Energy -aware Data Delivery Framework for Safety-Oriented Mobile IoT," *IEEE Sensors Journal*, vol. 18, no. 1, pp. 470–478, 2017.
3. X. Baro, S. Escalera, J. Vitria, O. Pujol, and P. Radeva, "Traffic Sign Recognition using Evolutionary Adaboost Detection and Forest-ECOC Classification," *IEEE Transactions on Intelligent Transportation Systems*, vol. 10 (March, 2009): 113–126.
4. J. van de Beek, M. Sandell, and P. Börjesson, "ML Estimation of Timing and Frequency Offset in OFDM Systems," *IEEE Transactions on Signal Processing*, vol. 45 (July, 1997): 1800–1805.
5. IEEE 802.16-REVd/D5-2004, "IEEE Standard for Local and Metropolitan Area Networks—part 16: Air Interface for Fixed Broadband Wireless Access Systems," May 13, 2004.
6. F. Al-Turjman, and A. Radwan, "Data Delivery in Wireless Multimedia Sensor Networks: Challenging & Defying in the IoT Era," *IEEE Wireless Communications Magazine*, vol. 24, no. 5, pp. 126–131, 2017.
7. S. Mumtaz, K. M. S Huq, A. Radwan, and J. Rodriquez, "Energy Efficient Scheduling in LTE-A D2D Communication," *IEEE Comsoc Multimedia Communications Technical Committee*, Vol. 9, No. 1 (January, 2014).
8. F. Al-Turjman, "Information-Centric Framework for the Internet of Things (IoT): Traffic Modelling & Optimization," *Elsevier Future Generation Computer Systems*, vol. 80, no. 1, pp. 63–75, 2017.
9. 3GPP TS 36.300, Evolved Universal Terrestrial Radio Access (EUTRA) and Evolved Universal Terrestrial Radio Access Network (E-UTRAN), Rel. 10, v10.2.0, December, 2010.
10. H. C. Hsieh and J. L. Chen, "Distributed Multi-Agent Scheme Support for Service Continuity in IMS-4G-Cloud Networks," *Computers and Electrical Engineering*, vol. 42 (February, 2015): 49–59.
11. H. Lee, T. Kwon, and D. H. Cho, "Extended-rtPS Algorithm for VoIP Services in IEEE 802.16 Systems." *Proceedings of the IEEE International Conference on Communications*, Istanbul, Turkey (June, 2006): 2060–2065.
12. F. Al-Turjman and H. Hassanein, "Towards Augmented Connectivity with Delay Constraints in WSN Federation", *International Journal of Ad Hoc and Ubiquitous Computing*, vol. 11, no. 2 (2012): 97–108.
13. 3GPP TS 23.167, IP Multimedia Subsystem (IMS) Emergency Sessions, Release 11, September, 2013.
14. 3GPP TS 23.228, IP Multimedia (IM) Subsystem Cx and Dx Interfaces; Signalling Flows and Message Contents, Release 11, September, 2013.
15. M. Patel, "SOS Uniform Resource Identifier (URI) Parameter for Marking of Session Initiation Protocol (SIP) Requests Related to Emergency Services," draft-patel-ecrit-sos-parameter-07.txt (October 26, 2009).

Chapter 8

Intelligent UAVs for Multimedia Delivery in Smart-Cities' Applications

Fadi Al-Turjman and Sinem Alturjman*

Contents

* Antalya Bilim University, Antalya, Turkey

143

8.1 Introduction

Advances in the Internet of Things (IoT) technologies and some new emerging ICT technologies such as the 5G devices/sensors are converging with a variety of application fields [1, 2]. Its integration with the industry is envisaged to revolutionize the current industry by creating smarter machines, building connectivity between them, and allowing them to communicate with and control one another for collaborative automation and intelligent optimization. 5G is expected to be more than a new generation of mobile communications [3]. It is already considered the unifying fabric that will connect billions of devices in some of the fastest, most reliable, and most efficient ways possible. Of course, the impact of such an enabling technology is expected to be revolutionary. The new infrastructure for communication is expected to transform the world of connected sensors and reshape industries. Such a revolution would of course require research and development for the coexistence and device interoperability of sensors with 5G networks.

Drones, also known as Unmanned Aerial Vehicles (UAVs), have been used mainly in military applications for many years. However, there has been a recent increase in the use of UAVs in non-military fields, which is inspired by the 5G revolution. Such fields include precision agriculture, security and surveillance, delivery of goods, and provisioned services [1]. For example, Amazon and Walmart have been working on a new system to deliver goods to customers through the air. Additionally, China's largest mailing company DHL, has started delivering around 500 parcels daily using UAVs. Moreover, we can use some 5G-supported UAVs to monitor and send feedback from incidents that happen along the road, hence, eliminating the need for road support teams. Moreover, a traffic policeman can be replaced or assisted by a UAV, by having it hovering over fast-moving vehicles and reporting back traffic violations. Consequently, the use of UAVs for industry-oriented services can become a reality very soon, especially after the revolution of communication systems toward realizing the 5G-inspired Internet of Things (5G/IoT) paradigm. A key field of interest for IoT and sensor networks is the development of wearables that can connect to these UAVs for various application areas. Having an infrastructure such as 5G which is developed considering IoT applications in detail causes significant need for further contributions in terms of data and especially multimedia delivery. 5G is targeting 10 Gbps data rates in real-time networks [4]. The recent tests in 5G technology have shown that it is even possible to exceed 1 Tbps in a laboratory environment. This would mean being able to transmit 33 HD films each up to 2.5–3 hours in a single second. It is typically desirable to have these infrastructures seamlessly integrated with IoT industrial solutions and there are recommended prototype sensors for similar applications with the energy and marginal cost for each added sensor [6].

Wireless Sensor Networks (WSNs) are very critical in the aforementioned archetype. The integrated 5G and IoT is an extraordinary complex model, where devices are deployed as consumer elements forming a complex interconnected system. Conversely, these elements operate within very strict energy constraints, and hence make the energy left over for fault-tolerance procedure very limited. Moreover, the emergence of the variety of multimedia IoT applications such as video streaming from smart homes in smart cities will certainly increase the need for a fault-tolerant data routing [7].

Nowadays, WSNs function in an autonomous manner with very limited human control in a UAV-enabled system, where sensors and cameras are attached/distributed not only in smart environments, but also to flying UAVs in the industry. Moreover, most of these sensors are positioned in wild outdoor environments and sometimes even harsh environments. Hence, it is quite difficult to determine and design a fault-tolerant routing protocol. Because the communication energy is considerably lower than that used in computations, it is very important to come up with a fault-tolerant algorithm which is able to recover from path failures no matter what the added computational energy is. Or else, any random event may cause the UAV failure in delivering their exchanged information and interrupt the network functionality.

Accordingly, this necessitates a multipath routing approach that can recover the failed path. Multipath routing protocols form a good candidate for a more reliable 5G/IoT paradigm, in which fault-tolerance routing problems are considered as optimization problems. These optimization problems formulate a k-disjoined paths to encounter up to $k-1$ path failure. Exceptional fault-tolerance routing in UAV-enabled networks needs huge computational power, which as the problem increases, brings about large control message overhead without scalability [8]. Coming up with a solution to these problems on each sensor may require significant capacities in terms of memory and computational resources, and still produce ordinary results.

To offer quick recovery from failures, we designed a bio-inspired routing algorithm called Particle Swarm Optimization (PSO). Authors in Jiang et al. [9] note that the use of PSO has produced positive results, due to its simple concept and high efficiency. Nonetheless, despite the competitive performance, there still remains a huge challenge of solving the fault-routing problem because of the convergence issue. However, many of the impulsive convergence traps occur due to fast convergence features and a diverse loss of particle swarm, and hence, result in different solutions. In addition, the ability to differentiate between exploration and exploitation search is another significant challenge that we face today. Exploration contains the swarm convergence, while exploitation usually tend to make the swarm particle convergence without leaving the viable area that eventually leads to premature convergence, hence it is never proper to overemphasize exploitation or exploration [11]. Due to these challenges that must be faced, and especially the connectivity issues, we propose a new approach that is more efficient in recovering failures via multipath routing capable of attaining Quality of Service (QoS) in terms of energy consumption, lifetime, delay, and throughput. The proposed multipath routing algorithm is compared against an existing optimization algorithm, namely Canonical Particle

Swarm (CPS) [12], Fully Multi-Path Swarm (FMPS) [10] and Canonical Particle Multi-Path Swarm (CPMS) optimization algorithm [11] to offer a different learning technique for the swarm particles. The aforementioned algorithms are only different from each other in that they have different learning contrivances and the likes, otherwise, they are similar to each other. Additionally, increasing the number of paths requires more messages exchange and communication overhead [12]. Therefore, we adopted the use of intricate network connections so as to denote layout of the swarm and use the multipath routing algorithm to stabilize the trade-off between fault tolerance and communication overheard by taking advantage of a mixed proactive and reactive routing mechanism that maintains the best objective function value for the designated paths per particle. After this, the particles are increased or decreased and then given a velocity that suits them where the augmented objective function must be used in order to make a fitting assortment.

In view of IoT solutions for the manufacturing industry from a system and network perspective, this study endeavors to provide novel data delivery solutions to gain machine/sensors interoperability and manufacturing flexibility through production line level machine collaborations focusing on: (1) sensor/machine functionality and decentralized structure for communication intensive applications; (2) ubiquitous message trading and learning techniques for collaborative automations; (3) Swarm-based management for application level flexibility and adaptation. Due to the aforementioned issues in WSN technology, and especially the connectivity ones, we propose a new routing algorithm that is more efficient in considering multipath failures which contain reconstructive procedures capable of attaining QoS in terms of network lifetime, energy consumption, delay, and throughput.

8.2 Related Work

Industrial Internet of Things (IIoT), also known as industrial internet, brings together smart machines, innovative analytics, and people at work [20]. It is an interconnection of many devices through a diverse communication system to bring forth a top notch system capable of monitoring, collecting, exchanging, analyzing, and delivering valuable information. These systems can then help manage smarter and faster business resolutions for industrial companies. This futuristic concept is marked with some new coined terms by industrial professionals and communities, such as Industrial 4.0, IIoT, Smart Manufacturing, Digital Manufacturing, Manufacturing 2.0, and Industrial Internet.

IIoT is more advanced than commercial IoT, simply due to the dominance of the connected sensors in the industrial platform [21]. Sensor interface is a key factor in industrial data collection, and the present connected number, the rate of sampling, and the type of signals emitted by sensors, are determined by the sensing device [13]. Additionally, every sensor connected to a device needs to write long and complex codes. Hence, Chi et al. [13] proposes a different system to design a programmable sensor interface for IIoT WSN, which is able to collect data and at the same time read it in real time from multiple sensors in high speed. There are many motivations associated with the IIoT such as connecting sensors to analytics and other data processing systems to automatically improve the industrial system performance, safety, reliability, and energy efficiency while collecting data using sensors, which have proved to be effective in terms of cost [14, 15]. The performance of machines and their connectivity, communication, and data throughput are all expected to be more flexible or powerful in manufacturing industry compared to traditional scenarios, such as smart home, health monitoring, and elderly care. The specificities of IIoT manufacturing industrial networks can be briefed as follows: (1) Powerful machines connected to sensors with reliable connectivity, (2) Real-time communication for collaborative automation, (3) Heterogeneous networked machines topology, (4) Arbitrary peer-2-peer (P2P) or broadcast for ubiquitous messaging, (5) High data throughput with varying sized messages using next generation paradigms (e.g., 5G), and (6) Machine collaboration based intelligent automation. These features have raised challenges and become the major concerns in the development of data routing in IIoT industrial systems.

In general, providing reliable services for applications which demand low latency within the 5G and the IoT context is a challenging issue. It is well known that some WSN industrial applications require deterministic systems with a reliable and low latency aggregation service guarantees. Since the IEEE 802.15.4e standard is considered as the backbone of the IoT regarding WSN, the existing Low Latency Deterministic Network (LLDN) model used to fulfill the major requirements of these applications require further contributions. In turn, research groups for this standard have studies on improvement of quality of service related concerns, including energy efficiency [16]. Apart from the fact that fault-tolerant routing makes the network system more reliable, it is also very important when it comes to 5G or IoT since WSNs heavily depend on surrounding environments and interact with them, and hence, the need to provide QoS is a must. There are different ways to determine a fault-tolerant route in WSNs, one of the leading ways is through multipath routing [17]. Adnan et al. claims that recent optimization methods, including meta-heuristic ones, are more effective in multipath routing [18]. Moreover, Al-Turjman comes up with a solution to the disjoined multipath problem, by proposing a new energy efficient multipath routing system based on using PSO [19]. Shieh et al. [23] introduces PSO, which is used to select routes for load delivery. On the other hand, Zhou et al. [24] recommends an Enhanced PSO-based Clustering Energy Optimization (EPSO-CEO) system that minimizes the use of power in each node by using centralized clusters and optimized cluster heads. Nevertheless, none of these studies address jointly the lifetime and fault-tolerance aspects in their routing algorithms with a convergent model.

Both Pant et al. [22] and Zhou et al. [24] present a model to prevent the unnecessary convergence of crowd by setting upper and lower search space bounds so as to enable the crowd to find solutions for diverse applications. Furthermore, Hasan et al. [25]

demonstrates the performance of a load distribution system so as to address the optimization problem and facilitate prime network selection. Moreover, the authors account for the network bandwidth and the errors in the ideal data delivery system in different networks with reduced cost. To improve the lifetime and increase the bandwidth of the energy proficient distributed clusters of sensors, Hasan et al. [26] employ the use of an energy-aware, delay-tolerant, and centralized approach. However, traditional sensor networks spend energy in almost all processes. They spend energy while making data transmission and data sensing as well as data processing. There have been a few attempts toward achieving more energy efficiency in such networks via wireless multi-hop networking such as in Al-Turjman and also Singh and Al-Turjman [27, 28]. However, such schemes are mostly applicable in static environments and can struggle with random topologies. For example, a routing scheme for energy harvesting was proposed in Hasan et al. [29]. The routing scheme assumes a hierarchical cluster-based architecture. Packet transmission from the source to the cluster-head can be direct or multi-hop based on the probability of saving energy through careful transmissions, optimized throughput, and minimized work load.

In this work, we propose a new routing algorithm that is more efficient in considering multipath failures which contain reconstructive procedures capable of attaining QoS in terms of network lifetime, energy consumption, delay, and throughput, while taking into consideration the advantageous powerful machines in IIoT industrial systems.

8.3 System Model

The routing method that has been projected in this research uses a fault-tolerant system in two-tiered heterogeneous WSNs which comprise super/smart nodes that have plenty of resources and simple/light sensor nodes with limited battery capacity and absolute QoS limitations. Nevertheless, to get a more resilient fault-tolerant network model, we look for a k-disjoined multipath routing approach. Furthermore, we look at a many to one traffic system, where super nodes and common nodes connect with the proper degree. Below, we list some important explanations of some terms before introducing the system model.

For every disjoined/isolated node, we have to use it to build a k-disjoint multipath route, and increase the number of marginal paths, and hence, render a fault-tolerant network. Our model is based on the assumption that a given node can connect or disconnect with some nodes that are not among those that are on the k-disjoint multipath between the node and a super node. In this study, node-disjointness relations are modeled as a directed graph $G(V, E)$, where $|V| = \{v_1, v_2, \ldots, v_N, v_{N+1}, \ldots, v_{N+M}\}$ is the fixed number of nodes or particles, N indicates sensor node while M signifies super/smart nodes, G represents the set of paths and the relationship between a pair of super nodes and a pair of particles is the number of edges E in G. $E = \{(v_i, v_j) | \text{Hop}(v_i, v_j) \leq \tau\}$, where $\text{Hop}(v_i, v_j)$ is the distance between v_i and v_j. $P(v_i, v_j)$ is a path that runs from v_i to v_j in graph G. it is a sequence of edges we get when we go from v_i to v_j where $i = j = 1, 2, \dots, N + M$. Hence, we can describe G as a set of unconventional routes $p_i(v_i, v_j)$. $e \in p_i(v_i, v_j), (v_N, v_{N+M})$ denotes a connection between any two nodes, $E(v_i, v_j \in p_i(v_i, v_j))$ is the node disjoint between $p_i(v_i, v_j), (v_N, v_{N+M})$ and e. Hence, we can obtain k-disjoined paths in G. We use the amount of energy consumed by a multipath, delay, and throughput to evaluate how best a multipath performs, we use the roots on Hasan et al. [25] and Hasan et al. [26] to come up with a solution of the objective function that minimizes the energy consumption and average delay, and maximizes the system throughput and network lifetime.

8.3.1 Problem Formulation

For our problem statement, we are looking to design a k-disjoined multipath for a fault-tolerant system, which uses a UAV to transmit multimedia to a super node located in a two-tiered WSN. The model is constructed in such a way that each sensor node in the network is within the transmission range of each other. This helps in minimizing the QoS parameters such as the transmission power level and latency while maintaining a k-disjoint multipath route. In this system, every sensor node is connected to at least one super node with k-disjoint multipath. Accordingly, a k-disjoint multipath constructed by linking a cluster of super nodes (or UAVs) with a bunch of sensors which can modify their transmission range to a prime value. The transmission range of each sensor should be such that the minimum amount of energy is used while still maintaining a k-disjoined multipath and all the parameters for QoS are still upheld.

8.3.2 Energy Model

For proper energy limitations, we need to consider the number of hops and the distance between two UAVs along the predefined path. The neighborhood topology mentioned in Section 8.3 is used. Each sensor node is within the transmission range of its neighbor sensor node. Given that the transmission range is equal to t_u (>0), then the neighborhood is formulated by:

$$\aleph_{u,v} = \left\{ v, u \neq v \, | \, \|n_u - n_v\| \leq t_u \right\} \tag{8.1}$$

It is worth recognizing that there is a chance that this might change during the dynamic network lifetime. Moreover, unless all constraints are met, there will be a division in the multipath and the neighborhood will be reconstructed. We can use the constraints to change the topology of the system which will eventually lead to solving the optimal power problem. Hasan et al. [25] uses the method of cut-off value to determine the lower and upper bound of the number of hops and transmission range. E_{elec} represents the energy derived when using the transmitter and receiver circuitry. The energy used by the receiver to obtain a proper signal to noise ratio is represented by ε_{mp}. The amount of energy loss during transmission is α, and $\tau_{n(nu,nv)}$ is the transmission range. To conclude, the function used to get the minimum amount of energy used in one node to transport data of length L_p for a distant of τ is formulated by:

$$\min \vec{Z}$$

S. t.

$$\text{hop} = \sqrt{\alpha \tau_n \left(n_u, n_v \right) \left(\frac{3\varepsilon_{mp}}{2 E_{elec_{n(n_u,n_v)}}} \right)} \leq \tau_n \left(n_u, n_v \right) \qquad (8.2)$$

$$\vec{Z} = \text{Energy}_{n_{sd}} = L_p \left\{ \sum_{n_n}^{n_d} 2 \left[E_{elecn_{sd}} + \varepsilon_{mp} \left(n_{sd} \right)^\alpha \right] \right\} \qquad (8.3)$$

The above energy value for the selected path can change according to the selected upper/lower bounds, E_{min}, and E_{max}, respectively. It represents the minimum and the maximum constants, respectively.

8.3.3 Delay Model

In this research, we consider the delay definition which depends on the hop count, denoted as $\varphi(\xi_i, \xi_j)$. φ represents the delay between two nodes, its definition is determined by the ideal number of hops. Given the optimal number of hops in Equation (8.3), which represents the minimum delay between two nodes, we can formulae and optimize the route selection while considering delay and network resources constraints. This optimization problem, shall consider both source and intermediate nodes periodically in the immediate neighborhood. Additionally, if one sensor node gratifies one QoS, the problem converges and all QoS requirements will be achieved. Consequently, the end-to-end delay for a given path P between is φ_{Source} and ξ_{Sink} is described as:

$$\varphi_{sourcessink} \left(L_p \right) = \min \left\{ \sum_{\xi_i} \varphi \left(\xi_i, \xi_j \right) \right\}, \qquad (8.4)$$

where $\varphi_{SourcesSink}$ denotes the minimum delay that we can achieve when we send data through paths between ξ_{Source} and ξ_{Sink}. This time consists of the time for transmission, retransmission, staying idle, queuing, propagation, and processing. And thus, considering

$$\sum_{v=1}\varphi\left(\xi_i,\xi_j\right)\le X_v\Delta_\varphi \tag{8.5}$$

the average delay per sensor node is equal to ξ. Assuming the hop count on a path between ξ_{Source} and ξ_{Sink} is given by η_{ij} and the delay along this path is L_e^φ. The hop delay constraint can be signified by $L_e^\varphi = \dfrac{\Delta_\varphi - \varphi^e}{\eta_{ij}}$. Accordingly, we can rewrite the constraint in Equation (8.5) as

$$\sum_{u=1}\varphi\left(\xi_i,\xi_j\right)\le X_u L_e^\varphi \tag{8.6}$$

8.3.4 Throughput Model

According to reference [20] a definition of throughput can be used to represent the number of data packets successfully transmitted. This can help in calculating the optimal hop count while maximizing the network throughput. This Throughput (Th) is computed via the following:

$$\text{Th} = \left(\frac{L_e^\varphi}{\xi}\right) * \text{TR} \tag{8.7}$$

where TR is the Transmission Rate.

8.4 Particle Optimization in IIoT

In view of the technical solutions for IIoT systems, it is necessary to describe the components and their interrelationship, namely, the roles of machines, and how they generate, exchange, and consume data to fulfill the interactive operations. From a Swarm system perspective, industrial machines/sensors can be considered as particles with unique associated characteristics/events.

Every sensor/particle is allocated to a *k*-disjoint multipath according to the sensor transmission level and required separation distance per hop as alluded to in Equation (8.3). This is performed while considering numerous swarm particles' attributes. In this approach, sensor nodes have the capacity to enhance agreeable learning conduct by trading path-related messages with their neighbors. After trading/exchanging these messages, every node/particle configures the disjointed ways and expands the neighborhood set as alluded to in Equation (8.3). As indicated by Equation (8.2) and Equation (8.4), another potential set of paths will be formed and prioritized, and thus, hops per the *k*-disjoint multipath will be adaptively changed according to the particle speed $v_{(i,j)}$ that is refreshed after each iteration to fulfill the desired QoS requirements.

Given that a k-disjoined multipath can have m descriptive QoS attributes, the position and velocity of the particle v is given by an m dimensional vector $|V| = \{v_1, v_2,...,v_N, v_{N+1},...,v_{N+M}\}$. The proposed Swarm algorithm in this chapter contains p_{best} and g_{best} which are the personal and global best positions respectively. By solving Equation (8.2) and Equation (8.4) in terms of the average consumed energy and average delay, we find the nodes that connect the entire searching space in every iteration. Trading control messages between the nodes can further trigger them, which are then defined as extreme value and global extreme value. Consequently, this leads to the extreme value within the feasible search space toward which we progress upon after every iteration. Hence, the nodes that tend to diverge are excluded. The personal best position of the swarm is brought about by the dissemination of good objective functions as denoted in Equation (8.2) and Equation (8.4). These equations are concerned with the information exchange, while satisfying the constraints which are used to get the velocity and then find the ideal multipath route as stated before. In every path, the personal best position of a particle $v_{(i,j)}$ is given by $p_{best,v(i,j)} = (p_{(best,v1),(best,v2),\cdots,(best,vN),(best,vN+M)})$, similarly, the global best position of a particle is given by $g_{best,v(i,j)} = (g_{(best,v1),(best,v2),\cdots,(best,vN),(best,vN+M)})$. The extent to which the $p_{best,v(i,j)}$ affects the equation is given by the coefficient of constraints φ_1, similarly, the effects of the global are denoted by the coefficient of constraints φ_2. The velocity of the updated function drives what we call the CPS optimization, where $\vec{Z}$ represents the distribution of the objective function, found after satisfying the constraints mentioned above. x is the constriction coefficient which assists in balancing global and local probes. It is represented as $x = 2 / \varnothing + \sqrt{\varnothing^2 - 4\varnothing}$, where $\varnothing = \varnothing_1 + \varnothing_2 > 1$. Equation (8.6a) is defined as the velocity update function and regarded as the momentum function, which gives the particle's/node's present direction. Equation (8.6b) is called social component, it has the ability of being drawn toward the best solutions as assessed by the neighbors. Equation (8.6c) represents the cognitive module, with the ability of being drawn toward earlier results which symbolize the node behavior. The only difference between CPS and the Fully Particle Multi-path Swarm Optimization (FMPS) algorithm is the function used to update the particle velocity. This means that we not only take into account the best position of the node, but also that of all of its neighbors.

We can ignore some of the node fault-tolerance messages which can lead to trapping in local optimal solutions, by eliminating the exchanged information about the personal-best $p_{best,v(i,j)}$. Consequently, this can lead to the increase of the node's ability to learn from the experience of other nodes. Hence, the performance of the algorithm highly rests on influence of the nodes while satisfying the objective function. Algorithm 1 denotes the pseudo code of the proposed CPS algorithm. It finds the p_{best}'s objective function given by Equation (8.2) and Equation (8.4) first in terms of the consumed energy and average delay, then finds the least value of objective function in the p_{best}'s objective function for k-disjoined multipath. Then, it assists in avoiding velocity fit and computes the constriction value x, as shown in $\vec{v}_p$. It updates the velocity value, and finally, establishes a better fault-tolerant multipath route on which the opal nodes are chosen.

ALGORITHM 1: CPS

1. input: Objective functions $f(x)$;

2. $X := \{x_1, \ldots, x_n\} := \text{InitParticle}\left(\overrightarrow{lb}, \overrightarrow{ub}\right) \to \forall_p \in \{1, \ldots, n\} : \vec{x}_n \vec{\psi} \vec{U}\left(\overrightarrow{lb}, \overrightarrow{ub}\right)$

3. $V := \{v_1, \ldots, v_n\} := \text{InitParticleVelocities}\left(\overrightarrow{lb}, \overrightarrow{ub}\right) \to \forall_p \in \{1, \ldots, n\} :$

$$\vec{v}_n \vec{\psi}\left(\overrightarrow{lb} - \overrightarrow{ub}\right) \otimes \vec{U}(0,1) - \frac{1}{2}\left(\overrightarrow{ub}, \overrightarrow{lb}\right)$$

4. $Y := \{\vec{y}_1, \ldots, \vec{y}_n\} := \text{EvaluateObjectfunction}(X) \to \forall_p \in \{1, \ldots, n\} :$

$$y_n \vec{\psi} f\left(\vec{x}_p\right)$$

5. $P := \{\vec{p}_1, \ldots, \vec{p}_n\} := \text{Initllocallocallyoptimal}(X) \to X$

6. $P := \{p_1^f, \ldots, p_n^f\} := \text{InitObjeectivefunction}(Y) \to Y$

7. $G := \{\vec{g}_1, \ldots, \vec{g}_n\} := \text{Initgloballyoptimal}(P, T) \to P$

8. $G := \{g_1^f, \ldots, g_n^f\} := \text{Initgloballyoptimal}\left(P^f, T\right) \to P^f$

9. **while** termination condition nor met do **do**

10. **for** each particle node p of n do **do**

11. $u_p = x^*\left(\vec{u}_p + \Delta(0, \varphi_2) \otimes \left(\overrightarrow{\text{locallyoptimal}_p} - \vec{x}_p\right)\right.$

$$\left. + \Delta(0, \varphi_2) \otimes \left(\overrightarrow{\text{locallyoptimal}_p} - \vec{x}_p\right)\right)$$

12. $\vec{x}_p := \vec{x}_p + \vec{v}_p$

13. **end for**

14. $Y := \text{EvaluateObjectivefunction}(X, f)$

15. $P, P^f := \text{Updatelocallyoptimal}(X, Y) \to \forall_p \in \{1, \ldots, n\} : \overrightarrow{p}_p, p_p^f$

$$:= \begin{cases} \overrightarrow{x_p}, y_i & \text{if } y_i \text{ better than } p_p^f \\ \overrightarrow{p}_p, p_p^f & \text{otherwise} \end{cases}$$

16. $G, G^f := \text{Updategloballyoptimal}\left(P, P^f, T\right) \to \forall_p \in \{1, \ldots, n\} : \overrightarrow{g}_p, g_p^f$

$$:= \text{best}\left(P_{T_p}, P_{T_p}^f\right), \text{ where } T_p \text{ are the neighbors of } p$$

17. **End while**

8.5 Performance Evaluation

Keeping in mind the end goal to evaluate the execution of the proposed Swarm method, we perform broad reenactments. We have implemented the aforementioned algorithms; the CPMS optimization, the CPS optimization, and the FMPS optimization, using MATLAB® in order to evaluate their objective functions and visualize their outputs. We use 100 sensors and 50 UAVs dispersed uniformly in $1000 \times 1000 \times 100$ m³ deployment space. The path loss exponent for the wireless communication model is chosen to be 2. The underlying estimation of sensors' transmission range is set to be 100 m in order to assure the association among UAVs and sensors nodes while fulfilling the focused on QoS requirements. Further simulation parameters are compacted in Table 8.1.

8.5.1 Simulation Results

Figure 8.1 presents the total energy consumption in the assumed topology with a maximum of 5-hops path length. We remark that the aggregate energy utilization in the k-disjoint multipath created by the proposed PMSO approach is superior to CPS. Since settling the objective function used by CPS experiences issues in finding k-disjoint multipath after recouping from operational failures in the immense search space. Furthermore, since this outcome in being not able to substitute the arranged multipath with some other options, more energy usage is experienced.

Another critical comment that is identified by traded messages for adaptation to failure between the super nodes (UAVs) and sensors is that CPS performs essentially even worse than FMPS and CPMS. This is because CPS necessitates

Table 8.1   Assumed Parameters

Parameter	Value
Message payload	64 bytes
Data length p	2000 bits
Transmission range	12.00 m
Tx data rate	250 kbps
E_{elec}	50 nJ/bit
Total number of UAVs	50 sensor nodes
ε_{mp}	0.0013 pJ/bitm²
Topology structure	Square (1000 m *1000 m)
Efs	10 pJ/bitm²

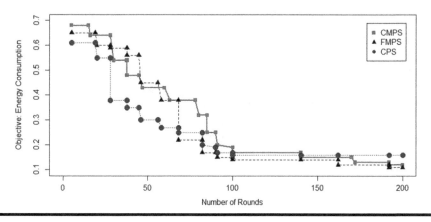

Figure 8.1   The swarm optimization routing in terms of energy consumption.

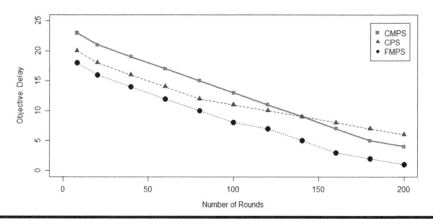

Figure 8.2   The swarm optimization routing versions versus average delay.

fundamentally more control packets to trade among the neighbors. In this way, CPS needs to discover k-disjoint multipath in its nearby neighborhood though the FMPS and CPMS can straightforwardly scan for routes utilizing less control packets among the nearby hops. Subsequently, the k-disjoint multipath for FMPS and CPMS can bring down the aggregate energy usage when contrasted with the CPS approach.

Figure 8.2 demonstrates the average deferral of the chosen ideal k-disjoint multipath from sender to the destination. We can watch that the assessed approaches; FMPS and CPMS, have exhibited a lower delay for every hop contrasted with CPS. This comes back to the determination and upkeep of the k-disjoint multipath for adaptation to failure which can fulfill the hop necessities by choosing the following bounce in the area of every hop. Therefore, it necessitates fundamentally less control messages for adaptation to non-critical failure contrasted with CPS for choosing and keeping up a 1-hop neighborhood. Hence, we can say that both CPMS and FMPS are more practical than CPS.

Throughput might vary because of high-BER and other surrounding changing conditions outdoors. Accordingly, in Figure 8.3 we show the impact of tackling the objective function alluded to in Equation (8.4). We notice that while expanding the ideal number of hops and trying to minimize the average delay, throughput degrades essentially. This is a normal influence for the assumed limited latency under the previously mentioned requirements while reducing the number of traded control messages. Therefore, the most practical hops on the route can be acquired effectively. Moreover, we presumed that the experienced algorithm's behavior relies upon the IIoT network topology. Despite the fact that CPS accomplishes completely connected topology, it has shown an exceptional degraded performance in contrast to other approaches. This is due to irregular conduct from every node while determining and limiting the multipath route hop count. At the same time, this conduct could bolster ideal execution in FMPS and CPMS with completely associated topologies.

We additionally explore the targeted performance when led with changing sensors/machines counts in the considered IIoT network topology. Results that have been created for every approach while expanding the quantity of sensor counts are shown in Figures 8.4, 8.5, and 8.6. It is worth pointing out that for multi-objective functions, the examined algorithms show better results at the beginning. Beside that the low connectivity degree (i.e., $k = 4$) could cause the algorithm to make mistakes in generating optimal approximations for the objective function. However, the behavior of particles while experiencing lower sensor counts can be utilized in estimating the objective function robustness, which could cause the algorithm to move toward more favorable regions in the feasible search space. It is observed that the performance for each approach is not good at the beginning

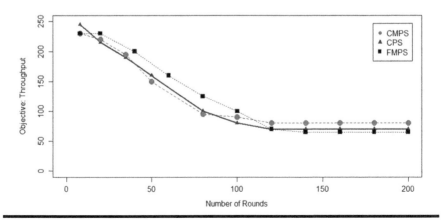

Figure 8.3   The swarm optimization routing versions versus throughput.

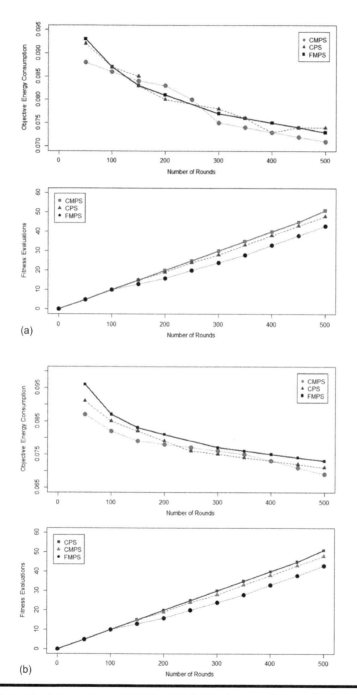

Figure 8.4 Total energy consumption with varying number of deployed nodes with different transmission range. (a) Energy consumption for 30 deployed sensors. (b) Energy consumption of 40 deployed nodes. (c) Energy consumption for 50 deployed nodes.

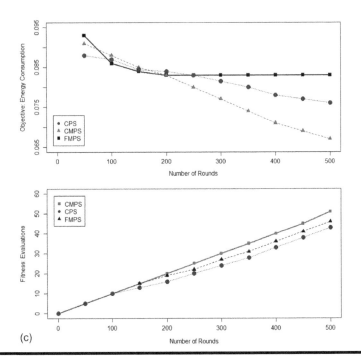

(c)

Figure 8.4 (CONTINUED)

but performs better toward the end while 40 sensors or more coexist as depicted in Figures 8.4b, 8.5(b), and 8.6b as well as for the 50 nodes as shown in Figures 8.4c, 8.5(c) and 8.6c.

According to these figures, FMPS and CPMS algorithms achieve the best performance when 30 coexisting nodes are assumed. Meanwhile, CPS performance is shown in Figures 8.4a, 8.5(a), and 8.6a in terms of energy consumption, delay, and throughput, respectively, and has the worst performance when the number of nodes deployed is more than 30 at the beginning and becomes better over time. This is because the CPS can construct and select optimal paths from an unfavorable area in the search space. Specifically, the low number of generation of paths for the objective functions can be a reason why the CPS's convergence is a little bit off from the global optimal solution for a 30 node deployment than with other algorithms.

Figure 8.7 compares the lifetimes of different counts of partitioning nodes, where a partitioning node is a node that can cause isolation/separation for a set of nodes in the network. In this figure, we consider a single data source (or UAV) and the network lifetime definition in 0. Accordingly, lifetime should be proportional to the ratio of the total deployed nodes' count N.

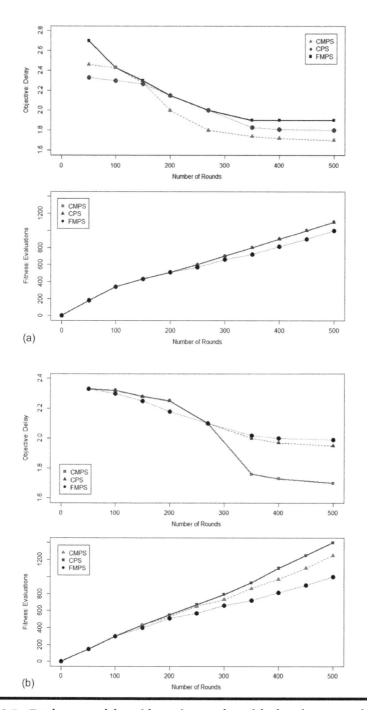

Figure 8.5   Total average delay with varying number of deployed sensor nodes with different transmission range. (a) Total throughput for 30 deployed nodes. (b) Total throughput for 40 deployed nodes. (c) Total throughput for 50 deployed nodes.

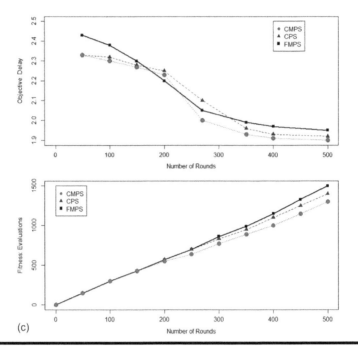

(c)

Figure 8.5   (CONTINUED)

Figure 8.8 illustrates the network lifetime for multiple data sources (or UAVs) and same setups in Figure 8.7. The only difference is that continuous multimedia traffic is transmitted by multiple sources to the partitioning nodes. This assumption makes the performance of the proposed Swarm-based algorithm easier to assess. Obviously, the network lifetime must be longer than that in the single source scenario. This is because the network lifetime is relatively proportional to the ratio of the partitioning nodes to UAVs' count per region in the network. It is worth pointing out here that with the same count of partitioning nodes, the network lifetime decreases when more than one source (UAV) is transmitting as depicted in Figure 8.8. When more UAVs are involved in covering a region, more energy is consumed by the network per time unit. Therefore, the lifetime is expected to decrease when the number of sources increases. Similarly, in the single source scenario (see Figure 8.7), the lifetime decreases when the partitioning nodes count increases.

8.6 Conclusion

IIoT is emerging as a dominant communication paradigm, nowadays, in order to satisfy the industrial revolution worldwide. In this research, we offer a bio-inspired swarm algorithm that constructs, recovers, and finds k-disjoint multipath routes in

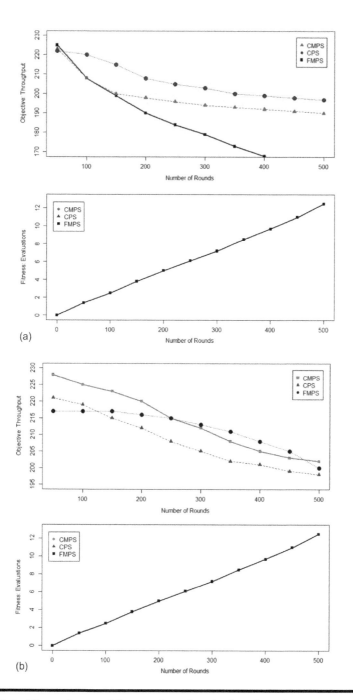

Figure 8.6 Throughput while varying the deployed nodes count. (a) total throughput for 30 deployed nodes. (b) total throughput for 40 deployed nodes. (c) total throughput for 50 deployed nodes.

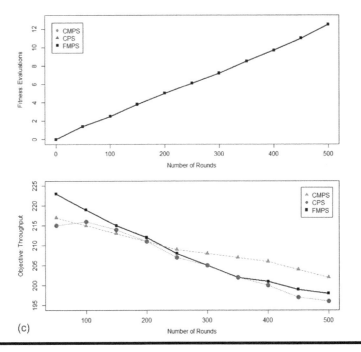

Figure 8.6 (CONTINUED)

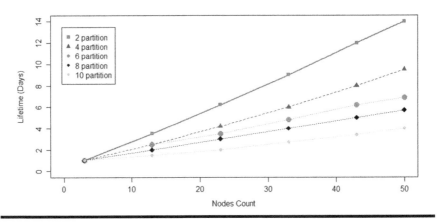

Figure 8.7 Node lifetime versus the nodes' count.

a network of machines (or UAVs). Two position information, namely the personal-best position and the global position, are considered in the form of velocity updates to enhance the performance of the IIoT network. In order to validate this algorithm, we assessed the multiple objective functions which consider throughput, average energy consumption, and average end-to-end delay. Our results show that

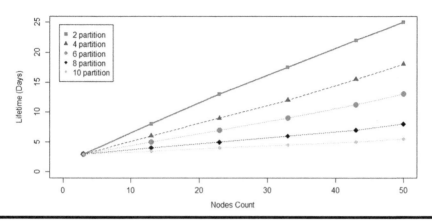

Figure 8.8   Node lifetime versus the nodes' count.

using the characteristics of all personal-best information is a valid strategy for the purposes of improving the CPMS performance. Moreover, the proposed algorithm has also been compared with similar algorithms, which optimize the energy consumption and average delay on the explored paths toward the destination. For the future, we see great potential in and a need to study various aspects of 5G/IoT integration with the existing sensor network architectures in different levels for more successful industrial applications. The popularity of 5G, the problem of slicing the Internet traffic, and the fact that a significant slice is expected to be reserved for sensory applications encourage further attempts in this domain.

References

1. GPP LTE version 15, [Online]: http://www.3gpp.org/release-15. Accessed on April 25, 2018.
2. F. Al-Turjman, "Energy -aware Data Delivery Framework for Safety-Oriented Mobile IoT", *IEEE Sensors Journal*, vol. 18, no. 1, pp. 470–478, 2017.
3. F. Al-Turjman, "Fog-based Caching in Software-Defined Information-Centric Networks," *Elsevier Computers & Electrical Engineering Journal*, vol. 69, no. 1, pp. 54–67, 2018.
4. M. Agiwal, A. Roy, and N. Saxena. "Next Generation 5G Wireless Networks: A Comprehensive Survey." *IEEE Communications Surveys & Tutorials*, vol. 18, no. 3 (2016): 1617–1655.
5. F. Al-Turjman, "5G-enabled Devices and Smart-Spaces in Social-IoT: An Overview", *Elsevier Future Generation Computer Systems*, 2017. DOI: 10.1016/j.future.2017.11.035
6. S. Elisa, S. D. Pascoli, and G. Iannaccone, "Low-Power Wearable ECG Monitoring System for Multiple-Patient Remote Monitoring," *IEEE Sensors Journal*, vol. 16, no. 13 (2016): 5452–5462.

7. F. Al-Turjman, and S. Alturjman, "Context-sensitive Access in Industrial Internet of Things (IIoT) Healthcare Applications", *IEEE Transactions on Industrial Informatics*, vol. 14, no. 6, pp. 2736–2744, 2018.

8. F. Al-Turjman, "Mobile Couriers' Selection for the Smart-grid in Smart cities' Pervasive Sensing", *Elsevier Future Generation Computer Systems*, vol. 82, no. 1, pp. 327–341, 2018.

9. S. Jiang, Z. Zhao, S. Mou, Z. Wu, and Y. Luo, "Linear Decision Fusion under the Control of Constrained PSO for WSNs," *International Journal of Distributed Sensor Networks*, vol. 8, no. 1, (2012): 871, 596.

10. Lim, W. H. and N. A. Mat Isa, "Particle Swarm Optimization with Adaptive Time-Varying Topology Connectivity," *Applied Soft Computing*, vol. 24 (2014): 623–642.

11. Wu, C. H. and Y. C. Chung, "Heterogeneous Wireless Sensor Network Deployment and Topology Control Based on Irregular Sensor Model," *Proceedings in Advances in Grid and Pervasive Computing*, C. Cérin and K. C. Li, Editors, Berlin Heidelberg: Springer, 2007: 78–88.

12. Vis, J. K., "*Particle Swarm Optimizer for Finding Robust Optima*", Leiden, The Netherlands, http://www.liacs.nl/assets/Bachelorscripties/2009-12JonathanVis.pdf, January 15, 2015.

13. Q. Chi, H. Yan, C. Zhang, Z. Pang, and L. Xu. "A Reconfigurable Smart Sensor Interface for Industrial WSN in IoT Environment." *IEEE Transactions on Industrial Informatics*, vol. 10, no. 2 (2014): 1417–1425.

14. B. Karschnia. "Industrial Internet of Things (IIoT) Benefits, Examples|Control Engineering," Controleng.com, 2017. [Online]. Available: http://www.controleng.com/single-article/industrial-internet-of-things-iiot-benefits-examples/a2fdb5aced1d779991d91ec3066cff40.html [Accessed: August 31, 2017].

15. F. Al-Turjman, "Price-Based Data Delivery Framework for Dynamic and Pervasive IoT," *Elsevier Pervasive and Mobile Computing Journal*, vol. 42 (2017): 299–316.

16. Y. Al-Nidawi, H. Yahya, and A. H. Kemp. "Tackling Mobility in Low Latency Deterministic Multihop ieee 802.15. 4e Sensor Network," *IEEE Sensors Journal*, vol. 16, no. 5 (2016): 1412–1427.

17. M. Dhir, "A Survey on Fault Tolerant Multipath Routing Protocols in Wireless Sensor Networks," *Global Journal of Computer Science and Technology*, vol. 15, no. 3, 2016.

18. M. Adnan, M. Razzaque, I. Ahmed, I. Isnin, Bio-Mimic Optimization Strategies in Wireless Sensor Networks: A Survey. *Sensors*, vol. 14, no. 1 (2014): 299–345.

19. F. Al-Turjman, "Information-Centric Sensor Networks for Cognitive IoT: An Overview," *Annals of Telecommunications*, vol. 72, no. 1 (2017): 3–18.

20. G. Singh, and F. Al-Turjman, "Learning Data Delivery Paths in QoI-Aware Information-Centric Sensor Networks," *IEEE Internet of Things Journal*, vol. 3, no. 4 (2016): 572–580.

21. F. Al-Turjman, H. Hassanein, and M. Ibnkahla, "Towards Prolonged Lifetime for Deployed WSNs in Outdoor Environment Monitoring," *Elsevier Ad Hoc Networks Journal*, vol. 24, no. A (January, 2015): 172–185.

22. M. Pant, T. Radha, and V. P. Singh, "A Simple Diversity Guided Particle Swarm Optimization," in *Proceedings of the IEEE Congress Evolutionary Computation*, (September, 2007): 3294–3299.

23. H. L. Shieh, C. C. Kuo, and C. M. Chiang, "Modified Particle Swarm Optimization Algorithm with Simulated Annealing Behavior and its Numerical Verification," *Applied Mathematics and Computing*, vol. 218, no. 8 (2011): 4365–4383.

24. Y. Zhou, X. Wang, T. Wang, B. Liu, and W. Sun, "Fault-Tolerant Multi-Path Routing Protocol For WSN Based on HEED," *International Journal of Sensors and Sensor Networks*, vol. 20, no. 1 (2016): 37–45.

25. F. Al-Turjman, "QoS -aware Data Delivery Framework for Safety-inspired Multimedia in Integrated Vehicular-IoT", *Elsevier Computer Communications Journal*, vol. 121, pp. 33–43, 2018.

26. S. Choudhury, and F. Al-Turjman, "Dominating Set Algorithms for Wireless Sensor Networks Survivability" *IEEE Access Journal*, vol. 6, no. 1, pp. 17527–17532, 2018.

27. F. Al-Turjman, "Cognitive Routing Protocol for Disaster-Inspired Internet of Things," *Elsevier Future Generation Computer Systems*, 2017. Doi:10.1016/j.future.2017.03.014.

28. G. Singh, and F. Al-Turjman, "Learning Data Delivery Paths in QoI-Aware Information-Centric Sensor Networks," *IEEE Internet of Things Journal*, vol. 3, no. 4 (2016): 572–580.

29. M. Z. Hasan, H. Al-Rizzo, and F. Al-Turjman, "A Survey on Multipath Routing Protocols for QoS Assurances in Real-Time Multimedia Wireless Sensor Networks," *IEEE Communications Surveys and Tutorials*, vol. 19, no. 3 (2017): 1424–1456.

Chapter 9

Intelligent IoT for Plant Phenotyping in Smart-cities' Agriculture

Fadi Al-Turjman and Sinem Alturjman*

Contents

---

* Antalya Bilim University, Antalya, Turkey

9.1 Introduction

Plant Phenotyping (PP) is the identification process of the genetic code differences and the environmental effects on the phenotype (or plant appearance/behavior). Phenotyping is a significant research direction in plant biological processes, and is used in both forward and reverse genetic approaches to obtain fundamental insights or advance crop improvement [1]. Phenotyping has been vital for many years where small scale farming, herding, and fishing were the principal means of existence for eras before the finding of oil and gas. Although the relative significance of these activities has decayed in recent years, governments are attempting to revive the animal and plant agriculture to provide a reasonable degree of self-support in food production. Moreover, they have recently issued a number of phenotyping projects such as medical phenotyping [1], Mediterranean fever in Turkish population [2], familial Mediterranean fever [3], and Alpha-Thalassemia Mutations [4]. Moreover, food supplies imports in Turkey, for example, recently decreased by 7.1 percent [5] while it can be much more effective when the Wireless Sensor Network (WSN) technology is utilized. Also worth mentioning is that the agricultural raw materials trading shortage is about $5 million [6]. In Qatar for example, they import over 90 percent of their food supplies, and the agricultural trading shortage is about US $1.2 billion [7]. Moreover, Qatar and many other similar countries are still providing less than 8 percent of their own poultry, 4 percent of their cattle, 7 percent of their sheep, and 6 percent of their own liquid fresh milk from domestic sources [7]. These numbers are expected to increase in the coming few years due to increasing population growth.

In general, the development of agriculture worldwide is facing major obstacles due to the scarcity of water, the problem of irrigation, use of fertilizers, seed breeding, machinery, pesticides, soil treatment, and soil analysis complications [6]. Therefore, reviving the agricultural sector requires the application of a novel farming model that has high economic efficiency, optimal use of scarce resources, minimum impact on the environment, and should be sustainable. One key strategic solution is to deploy Advanced Farming Management Systems (AFMS) to observe, measure, and respond to inter- and intra-field variability in the farms. PP arises as one of the key state-of-the-art technologies that can be adopted to build AFMS, which can introduce a significant increase to the crops, poultry, and livestock productivity by effectively managing the available resources and providing the optimum quantitates in terms of water, food, temperature, humidity, fertilizers, etc. The integration of PP will allow the farmers to establish comprehensive and informative records of their farms and crops, improve the decision-making process, and enable better management of their farms. Consequently, using PP can be one of the pivotal solutions for the aforementioned challenges. However, the process of taking phenotypic measurements had to be performed in laboratories under costly and time consuming conditions.

And so, in this survey article, we overview the existing attempts in WSNs toward realizing smart agriculture applications in general and PP in specific. We discuss key design factors in deploying WSNs under harsh operational conditions in outdoor environments. Moreover, we assess and evaluate the existing solutions and key enabling technologies that can realize the PP in practice in the near future.

The rest of this article is organized as follows. In Section 9.2, we discuss the PP project in detail in addition to providing some real examples from the field in Section 9.3. Next we introduce and discuss key design aspects related to PP testbeds in Section 9.4. In Section 9.5, we discuss the possibility of prototyping and implementing a custom WSN for PP in practice using the existing enabling technologies. Finally, we conclude this work with a few recommendations regarding possible future work in Section 9.6.

9.2 Background

The concept of PP, which was alternatively called site-specific management, is not new. The application of PP started more than 20 years ago [8]. The PP is generally associated with the application of crop production input elements based on the assessment of the variability of need for a particular input. The nature of the variability can be either temporal or spatial. The inputs are usually the regular daily requirements of the farms such as seeds, fertilizers, insecticides, drainage tiles, and subsoiling. The input can vary by rate (quantity) or type (e.g., two kinds of herbicide). PP suggests that the application of the input does not have to be uniform across the field, but the analyses of variability are essential to enable such an application. Based on this definition of PP, a grower who spot-sprays a field for weed control, or alters fertilizer quantities for a sandy knoll in his fields, is applying PP. Scheduling of irrigation, fertigation, and banding of fertilizers can be classified as PP practices as well. More recently, PP has progressed to include the use of advanced technologies such as Global Positioning Systems (GPS), yield monitors, field mapping and record archiving, variable rate application, and planting equipment. It is worth noting that PP should not be understood as the incorporation of advanced automated equipment in agriculture, but the gathering and effective use of information obtained from the field. Information collection and exploitation is the core of PP. Therefore, PP will not replace humans, but it will increase the capability and requirement for more highly trained farmers and engineers. Relatively speaking, very few groups of farmers in the world presently have such proficiency.. The PP can also be applied to livestock; however, it is commonly called Precision Livestock Farming (PLF). PLF enables the recognition of each individual animal. Using PLF has enabled farmers to record several aspects of each animal, such as age, pedigree, production, growth, health status, and feed conversion. The result is significantly higher reproduction outcomes, high-quality food and general safety, animal farming that is highly efficient and sustainable, healthy animals, and a low impact of livestock production on the environment [9].

9.3 Plant Phenotyping (PP) Examples

In Mutka and Bart [10], the authors summarize current progress in plant disease phenotyping and suggest future directions that will accelerate the development of resistant crop varieties. In Figure 9.1 we see a *Pseudomonas syringae* infection on *Arabidopsis thaliana* with gray water-soaked lesions surrounded by chlorosis. Figure 9.1b shows an early-stage *Xanthomonas euvesicatoria* infection on pepper with small water-soaked lesions. Figure 9.1c displays a *Xanthomonas oryzae pv. oryzae* infection on rice with grayish green water-soaked lesions coalescing into yellow streaks. Figure 9.1d shows a *Xanthomonas axonopodis pv. manihotis* infection on cassava with dark water-soaked lesions that are spreading and leading to leaf wilt. In Humplik et al. [11], the authors summarize as an example of the integrative automated high-throughput phenotyping platform, grow-chamber-based phenotyping (see Figure 9.2). This study focuses on recent advances toward development of integrative automated platforms for high-throughput PP that employ multiple

Figure 9.1 Examples of disease symptoms caused by bacterial plant pathogens discovered by phenotyping [10].

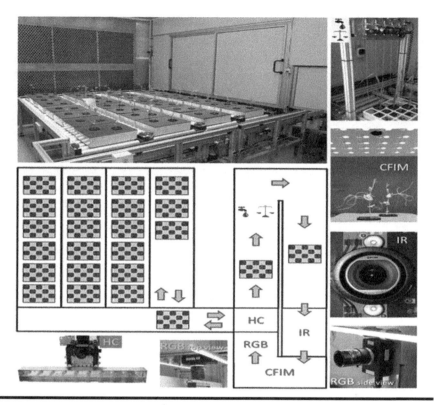

Figure 9.2   Scheme of grow-chamber-based automated high-throughput phenotyping platform [11].

sensors for simultaneous analysis of plant shoots. In both basic and applied science, the recently emerging approaches have found importance as tools in unraveling complex questions of plant growth, development, and responses to environment, as well as selection of appropriate genotypes in molecular breeding strategies. The development of effective field-based High-Throughput Phenotyping Platforms (HTPPs) remains a bottleneck for future breeding advances. However, progress in sensors, aeronautics, and high-performance computing is paving the way. In Araus and Cairn [12], the authors review recent advances in field HTPPs, which should combine at an affordable cost, high capacity for data recording, scoring and processing, and non-invasive remote sensing methods, together with automated environmental data collection (see Figure 9.3). Laboratory analyses of key plant parts may complement direct phenotyping under field conditions. Improvements in user-friendly data management together with a more powerful interpretation of results should increase the use of field HTPPs, therefore increasing the efficiency of crop genetic improvement to meet the needs of future generations. In Li et al. [13], the authors have assessed a range of different wavelength imaging techniques

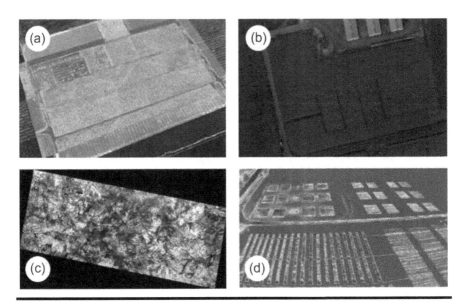

Figure 9.3 Different categories of imaging systems for remote sensing evaluation of vegetation are detailed below with examples of prototypes capable of being carried by UAPs of limited payload. Examples of false-color images taken with different categories of cameras: (a) RGB/CIR, (b) multispectral, (c) hyperspectral, and (d) thermal imaging [12].

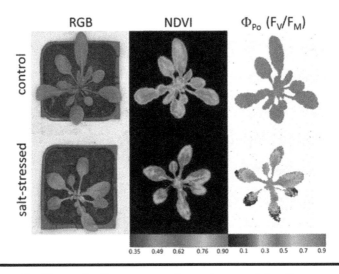

Figure 9.4 The illustrative figure presenting outcome of simultaneous analysis of control and salt-stressed Arabidopsis plants, using RGB, hyperspectral, and Chl fluorescence imaging [11].

in PP (see Figure 9.4). For the imaging sensors applied to PP, physical properties, depth knowledge, robust software, and image analysis pipelines are prerequisites to enable the collection of the phenotype data. These has been achieved by employing the wireless sensor network (WSN) technology. Fluorescence imaging was primarily used for foliar disease detection and thermal imaging for plant water status detection. A 3D surface reconstruction requires calibration for biomass estimation. Imaging spectroscopy requires standard procedures for the extraction of spectral features to reduce raw data in PP. There is a large difference in the reliability of imaging methods between controlled environments and the field. This reliability must be considered to understand the measurement principle for each experimental design, proper sensor calibration, and regular calibration of the imaging-based systems.

9.4 WSN Design Aspects in Phenotyping

The main objective of this chapter is to investigate design and implementation aspects of a WSN in phenotyping. The WSN shall handle remote monitoring and control of a large number of sensing and monitoring devices under variable density and mobility conditions. The network should operate under the constraints of low energy consumption and low power transmission, using reliable and cost effective networking techniques. The design should prolong the network lifetime and overcome various unpredictable elements at both network deployment and operation stages. Specifically aiming at PP applications, our interest is to enable features such as connectivity, coverage, reliability, survivability, and lifetime, in an energy efficient manner. In the following we propose a detailed discussion for the key design aspects to be considered in any PP application that involves a WSN as enabling technology.

9.4.1 Deployment

Deployment of nodes in the field is an important design aspect that is usually overlooked, or not thoroughly investigated, by designers of WSNs. Node placement can have a profound effect on connectivity, coverage, and reliability [13–15]. Therefore, deployment can be a significant factor in improving the cost-efficiency of the entire systems and the overall network performance. In the literature, most of the existing research on deployment tackles the problem from a 2D perspective [16, 17]. However, in PP applications the coverage area can change significantly over time in the 3D space, causing an optimized 2D deployment strategy to fail. For example, animal migration over seasons, plant growth in different directions, and different concentration of gases, are factors that can cause a significant change in channel propagation conditions. 3D deployment strategies can provide higher degrees of freedom for node placement thus leading to better connectivity and longer network lifetime. The main challenge of 3D deployment is that the search space is very large, thus optimization of node placement becomes computationally overwhelming. To address this issue, we have proposed a 3D grid-based deployment strategy [18], which limits the search space and simplifies the placement optimization problem. In this work, the monitored environment is divided into a virtual 3D grid, and the positions of Sensor Nodes (SNs) and Relay Nodes (RNs) are limited to the intersection points. We formulated an optimization problem that maximizes connectivity between the SNs and the Base Station (BS), with constraints on the required lifetime of the network, and the number of RNs available. Our proposed deployment strategy achieved outstanding performance results in terms of cost and network lifetime as compared to the 2D models. However, the work [18] was performed for low density static WSN where the nodes are placed in a deterministic manner, which is mainly suitable for small scale PP applications. In a previous work, we considered the 3D deployment problem taking into consideration the variable density, randomness, and mobility of the nodes [18]. This work optimizes the network connectivity, while guaranteeing specific network lifetime and cost. To evaluate the performance of WSN more accurately in a PP environment, a new metric shall be developed and utilized to measure the WSN lifetime. The revised definition for the network lifetime shall be developed by considering node redundancy and heterogeneity, in addition to addressing the problem of the huge search space and the network connectivity representation. The effectiveness of the deployment strategy shall be validated through extensive simulations and comparisons assuming practical considerations of signal propagation and connectivity under varying probabilities of node/link failures. Significant improvements are expected after considering a revised WSN lifetime definition in terms of the network connectivity and lifetime under harsh operational conditions where the probabilities of node/link failures are high.

It is worth noting that the deployment of the nodes shall be performed based on the recommendation of an expert (consultant) in the field of agriculture to optimize the nodes deployment process. For example, the number of nodes per unit area, and the variable sampling frequency has to be determined for targeted environment in particular. If such a process is not optimized, it can negatively impact the efficiency of the system.

9.4.2 Localization

In the more general PP models, the node density will be variable and thus varying Probability of Link Failures (PLF) might apply according to their location. Moreover, some of the nodes might be mounted on the farming machines, which also can change its location with respect to the other nodes. Mobility of the nodes can create variable density clusters of nodes and variable link quality. Detecting the location of the nodes can be used to maintain their connectivity to the network [19]. A few attempts in localization have focused on the development of efficient techniques to enhance the localization using wireless networks as well as satellite positioning systems [20–22]. For example, in Abdrabou and Zhuang [22] the authors presented novel technique for asynchronous WCDMA multipath delay estimation by decoding the received signal with a specific pulse shape, followed by Teager-Kaiser operator. The deconvolution process is applied to reduce the impact of the pulse shaping process utilized for band limiting purposes. In Jongerden and Haverkot [19], the authors considered Quality-of-Service (QoS) position-based routing for Ultra Wide Band (UWB) mobile ad hoc networks. The considered technique applies call admission control and temporary bandwidth reservation for the discovered routes through the network, taking into account the interactions of the Medium Access Control (MAC) protocol. Using cross-layer design, it exploits the advantages of UWB at the network layer by exploiting the location information for optimizing the routing and bandwidth allocation, and by enabling the multi-rate feature.

9.4.3 Medium Access Control (MAC) and Security

Energy waste at the link layer mainly comes from idle listening and overhearing [23]. In idle listening, a Mobile Node (MN) listens to an idle channel for possible incoming data traffic, while in overhearing the MN receives packets that are intended for other nodes due to the broadcast nature of radio transmission [24]. The common approach for energy efficiency in sensor networks is to put the MN into a sleep mode during its idle listening and/or overhearing time, i.e., to let the MN alternate its operation mode between active and sleep periods [25]. Extending such an approach to highly dynamic data traffic and node mobility in both temporal and spatial domains requires further studies, in order to satisfy QoS requirements [26]. Specifically, providing statistical delay guarantee for data delivery and satisfying requirements for packet dropping probability and throughput, while achieving energy saving, have been overlooked in literature. The situation is even more challenging in a multi-hop scenario, where it is necessary to avoid packet forwarding interruptions due to the next hop MN being in a sleep state. Focusing on distributed contention-based channel access, we recommend developing new methodology to effectively schedule the active and sleep periods so as to maximize energy efficiency while ensuring service quality in the presence of node mobility and data traffic dynamics. Intuitively, a longer sleep period for an MN saves more energy. However, it can jeopardize a desired upper bound on packet delivery delay at the destination. Due to the random nature of packet arrivals, wireless fading channel, and node mobility, a statistical delay guarantee can be more efficient than a deterministic guarantee. On the other hand, energy consumption can be reduced by decreasing MAC overhead and transmission collisions among nodes. One approach is to let a node contend only once to transmit a batch of packets within the required delay bound, followed by assigned contention-free channel time.

The sleep duration of an MN should be limited by its required throughput and the service demands of other MNs in the neighborhood. With distributed MAC, each node needs to contend for the channel according to incomplete information of its neighboring nodes, with possible asynchronous active/sleep schedules, via limited information exchange. The energy efficiency depends on MN active/sleep schedules, propagation environment, node locations, and traffic load. The source MN transmission shall be modeled as an M/G/1/K queue, given a Poisson data traffic arrival process, with a finite buffer size K to capture the delay bound requirement. Also establishing an analytical relation between the MN sleep duration and packet dropping probability with buffer size K, using the M/G/1/K queuing model is recommended [27]. The relation will help to capture the trade-off between energy efficiency and system throughput. Extension to a multi-hop scenario involves studying a G/G/1/K queuing system, which is technically very complex. Based on the single-hop analysis, we aim to develop some approximations for the energy and throughput relation.

Most existing works on energy efficient MAC protocols rely on time synchronization among MNs. However, it is not always practical to assume that MNs can satisfy the required time synchronization condition. We recommend developing a distributed asynchronous MAC protocol for a multi-hop, multi-source, and multi-destination network, based on cooperative information sharing, to minimize energy consumption at each MN while satisfying the required delay bound and throughput in PP.

The type of exchanged information in the WSN has different constraints and urgency in accordance with the content of the communicated packets. Thus, the way the security protocols are applied must match with the confidentiality required for that specific packet. This creates a need to classify different levels of communication before even relaying/broadcasting them at the WSN used in PP. The most crucial part in any WSN application nowadays is ensuring that the network supports an end-to-end encryption and authentication. Critical key points to be considered in small cells applications for guaranteed security and privacy protection are as follow: (1) personal data collection, which if limited to certain extent can significantly help in mitigating several security issues, (2) data and traffic analysis for WSN based applications requires information sharing, therefore service providers and the technology partners should come to an agreement on secure data handling methods which assures the mobile user privacy protections by considering the de-identification concept for example, (3) reliability of the WSN itself, encryptions and digital signature per user are also important aspects to be considered in this domain, (4) human errors, which can elevate security risks and breaches, and thus, customized policies and procedures are required to mitigate the oversight issues, (5) lastly, transparency of the WSN usage/configuration assures the integrity of such systems in wireless technology domains and necessitates accountably clear policies with respect to the offloaded data security and privacy.

9.4.4 Routing

Many new routing strategies have been proposed to solve routing problems in traditional WSNs [28, 29]. A Particle Swarm based routing approach for fault-tolerant optimization in heterogeneous WSNs which uses a hybrid routing scheme to calculate and maintain k-disjoint paths from source to Sink is proposed as well in [30]. It presents a model to solve the fault-tolerant optimization routing for SNs that are densely distributed in a heterogeneous wireless environment. It employs an intelligent swarm algorithm that provides a faster way to recover the k-disjoint multipath from failure. It is a cooperative algorithm, which defines each particle as an antibody to generate a new population in the space search. It provides accuracy in finding the optimal solution by jumping out to the local optima, while minimizing the energy consumption, and shorting the end-to-end delay in the packet delivery process.

Meanwhile, Hu et al. [31] have used a simple form of fault-tolerance mechanism, which relies on the directed connected graph concept. The construction of topological heterogeneous WSNs consists of two types of sensor equipment, arranged in two layers. The lower layer is formed by traditional SNs with restricted resources, which respond for any task, such as processing, transmission, and sensing data. The second layer consists of macro nodes with more capabilities in energy, processing, and storage and has the responsibility of decision routing. The design of routing protocols for PP, however, is still an open research area. In addition to the major issues of designing routing protocols in WSNs, there are new characteristics and constraints due to the nature of the energy-hungry multimedia exchange that must be handled over the network such that routing protocols for WSNs are not applicable to PP. Energy waste occurs mainly due to signaling overhead and unsuccessful transmissions during route discovery and repair. Routing protocols can be broadly classified into table-driven and on-demand protocols [28]. Table-driven routing is proactive, requires periodic advertisement of routing information, and is not considered to be energy efficient due to energy consumption associated with the routing overhead. On-demand routing is reactive, in which a transmission path is created only when needed. We will focus on on-demand routing, taking account of node mobility. Three main approaches exist to improve energy efficiency: (a) minimum energy routing which selects the most energy efficient path between the source and destination [29, 32], (b) data aggregation to concentrate the data traffic on some paths while switching off the MNs at the light loaded paths [33], and (c) energy balancing which aims at balancing the remaining battery energy at the MNs in establishing a route from the source to the destination [34, 35]. However, the first two approaches can exhaust the batteries of MNs along the selected paths. It can result in network partitioning which will eventually lead to a network failure. The third approach does not minimize energy consumption. We suggest developing energy efficient routing solutions that can balance energy saving with network lifetime maximization while providing a stable queuing behavior.

Also, we recommend more studies on cross-layer optimization for MAC and routing protocols. In a packet switching network, packets are often queued in transmitters for statistical multiplexing over a radio channel. While data aggregation routing can achieve energy saving, the queuing behavior of the network can be unstable if many paths are switched off and all data traffic is concentrated on only a few paths, due to high traffic intensity in the active paths. The queuing instability can have a severe detrimental impact on network performance and has not been studied. We will investigate the performance trade-off between the amount of energy saving (via data aggregation and path on/off switching) and the network stability. The alternating active/sleep modes at the link layer pose significant technical challenges in the queuing instability analysis. It is strongly recommended to address these challenges by using the statistical link layer model to be developed for PP, and by exploring conditions on the inter-arrival and service times of the queuing network [36]. This will shed some light on how many paths can be switched off and how to perform data aggregation so as to balance the amount of energy saving with the network stability. While data aggregation routing can reduce energy consumption, it can also reduce the overall network lifetime. Network lifetime can be the time to the first node failure, the time to the first network partition, or the time to unavailability of some application functionality [37]. Concentrating data routing on the minimum energy paths or data centric paths can exhaust the batteries of the MNs along the paths, which in turn reduces the network lifetime. Hence, an energy efficient routing protocol should balance energy saving with network lifetime maximization. Thus, we suggest first establishing an analytical model of the network lifetime as a function of MN residual energy, data traffic characteristics, and QoS requirements, based on MN energy models [38, 19] and graph theory [20]; second, to examine the energy consumption amounts for the minimum energy paths and for the data traffic centric paths, and evaluate performance trade-off between energy efficiency and network lifetime for the routing strategies; then, a routing protocol to balance energy efficiency with network lifetime maximization can be developed. The problem can be formulated as a multi-objective optimization, subject to QoS requirements and queuing stability constraints. We recommend using a weighting parameter to achieve the desired flexibility in the routing protocol, such as balancing energy saving and network lifetime or favoring one over the other. When each MN alternates its state between the active and sleep modes with the duty cycle adaptive to data traffic load and radio channel condition, the link layer exhibits a dynamic and random on/off behavior. A key issue to be addressed also is how to achieve a reliable end-to-end path under the link dynamics. One approach is to explore group routing, to route data packets through MN groups instead of employing a single forwarding MN [21]. If an MN along the most energy efficient path is in a sleep state, the MN's active neighbor(s) group can create an alternate suboptimal energy efficient path toward the destination. Further, MN transmit power is a control variable in the cross-layer design. By exploiting wireless broadcast nature in packet forwarding, we will develop a cooperative multi-hop relaying solution, which adaptively reorganizes the end-to-end path according to the instantaneous active/sleep modes of MNs, channel conditions, and MN locations to minimize the end-to-end outage probability while reducing the associated route establishment and maintenance cost.

9.4.5 Node-to-Node Energy Efficient Distributed Resource Allocation

The most substantial obstacle facing the deployment of small long-life SNs is the need for major reductions in energy consumption to maximize the energy saving of the system. An energy-aware design that highlights the elegant scalability of energy consumption with factors such as available resources and their significance, event frequency of occurrence, and desired output quality, at all levels of the system hierarchy. The design for energy-aware sensing nodes emphasizes the association between different layers of the sensing nodes stack to provide energy-quality trade-offs given that the hardware is designed for scalable energy consumption. The QoS in PP applications is mainly determined by the duty cycle of the data collection and the reliability of the collected information. In PP applications, the collected data will be varying in unpredictable manner. For example, a sensor for the soil moisture will not expect major changes over night in the absence of rainfall or irrigation. On the contrary, a rainfall or starting the irrigation system will cause a drastic change that has to be reported to the data processing center in real time to enable the irrigation control system to stop the irrigation process at certain times. Such behavior creates variable data traffic for the network and variable duty cycle for the sensing nodes. Therefore, an event-triggered transmission is necessary to maximize the energy efficiency of the nodes since the duty cycle for particular nodes can be as low as 5 percent.

One of the major limitations of the event-triggered WSN is that a large number of sensing nodes might be triggered simultaneously, which can cause network congestion, and it may overwhelm the data processing center with a massive amount of data to process, analyze, store, and interpret. However, a large part of the collected data could be correlated temporally and spatially, which results in transmitting a large amount of redundant information. Consequently, an optimum energy-aware WSN should consider the data correlation before even sending this data to the processing center. Toward this goal, it is necessary to enable node-to-node (n2n) communications so that the group of nodes in a given cluster are aware of the status of each other and hence they can select a small subgroup of nodes to transmit their data. At the data processing center, data fusion techniques can be utilized to create the desired field map. One possible approach is to borrow video processing techniques where particular nodes can be configured as base nodes and other nodes can be configured as enhancement nodes. The enhancement nodes can be classified with different enhancement levels. Then, we can use differential information measurements to select the frequency and level of nodes that will transmit. A simple example

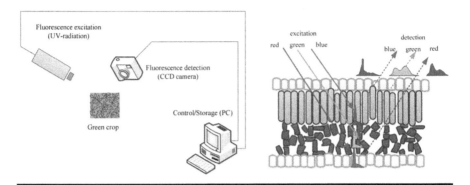

Figure 9.5 A scheme for the multi-color fluorescence imaging system and the chlorophyll fluorescence emission of green leaves as induced blue, red, and green excitation light [13].

of a node cluster that consists of 20 nodes is given in Figure 9.5. The base nodes' layer has four (red) nodes, the first enhancement layer (blue) has eight nodes and finally the second enhancement layers (green) consists of eight nodes as well. This structure has some similarity with the hierarchal modulation used in video transmission.

In the literature, the n2n communication is commonly referred to as Device-to-Device (D2D) communications. The D2D has received increased attention for wireless cellular networks because it can noticeably off-load the network traffic and reduce the transmission energy since the communicating devices are in close proximity of each other. Bluetooth is a simple example of D2D communications. However, the n2n approach and adopting the hierarchal topology requires mainly a specifically designed protocol because with n2n configuration the network becomes heterogeneous in the sense that it has distributed and centralized processing components. In Al-Turjman and Alturjman, and Ismail and Zhuang [39, 40], the authors considered the problem of distributed throughput optimization in wireless ZigBee networks as well as the problem of resource allocation in distributed heterogeneous wireless cellular networks. The results obtained revealed that the distributed topologies can be of significant impact on the network performance. However, the works in Al-Turjman and Alturjman, and Ismail and Zhuang [39, 40] were not designed under the same assumptions and conditions we are considering for PP applications. Hence, extending these algorithms to WSN for PP applications is not straightforward, and it will be one of the aims of any WSN applied in PP in the near future.

9.5 WSNs Prototyping and Implementation in PP

A typical wireless SN, depicted in Figure 9.6, consists of a power unit, processing unit, memory unit, sensing unit, and a communication unit. Each of these units has a strong impact on the overall performance of the WSN. The processing unit

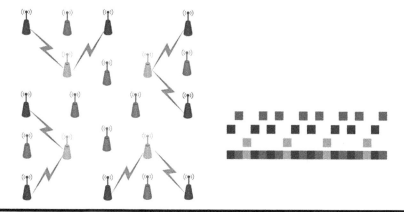

Figure 9.6 **Example of a hierarchical node configuration with three levels. The red nodes are the base nodes. The blue are the enhancement nodes level-1 and the green nodes are the enhancement nodes level-2.**

is the core component that collects the signals captured by the sensors, processes the gathered information, and transmits the information to the network via the communication unit. The processing unit has also access to a memory unit where the information can be stored if required. The communication unit allows communication with other nodes in the WSN using a wireless communication channel via a radio, infrared, or laser based link. The power unit provides the energy supply for the node and is composed of batteries, which may be rechargeable. The sensor unit is composed of multiple sensors and their choice depends on the application domain of WSNs. Typical sensors used for PP applications include soil moisture, relative humidity, temperature, and gas (carbon dioxide (CO_2), methane (CO_4), carbon monoxide (CO)) detection sensors [41]. Besides the above mentioned hardware modules, the most essential component of a WSN node is a dedicated microkernel. The operating systems used for other handheld devices cannot be used for WSN nodes because of their small size and rugged environments, and thus special attention has to be paid to problems like localization, filtering, energy consumption, and security. The most important implementation aspect of a WSN node is the choice of the processing unit, which greatly affects the energy consumption and the computational ability of a WSN node.

The typical choices are to use an off-the-shelf WSN node with a built-in processing unit, a microcontroller ,or a FPGA unit as the processing unit. Some of the commonly used off-the-shelf WSN nodes include Mica2, Mica2Dot, MicaZ, Fleck 3node, and TinyNode mote. These WSN nodes contain a microcontroller, which is used as the processing unit, along with most of the other blocks depicted in Figure 9.7 on the same board. The WSN can be setup by programming the microcontrollers and thus this choice provides the best time-to-market but on the downside it does not provide the flexibility to change

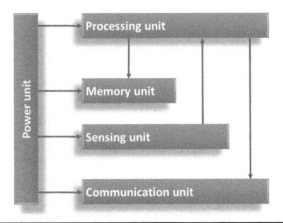

Figure 9.7   A high-level block diagram of a sensing node.

components. The other two choices, i.e., microcontroller and FPGA-based processing units overcome this problem and are discussed in detail in the following sections.

9.5.1 Microcontrollers

Nowadays, microcontrollers include their own memory (both volatile and non-volatile) units and various typically used standard modules like Analog-to-Digital converters, UART, timers, and counters. Thus, a WSN node can be constructed using a microcontroller by integrating such a microcontroller with a communicating unit and sensors. A wide range of microcontrollers dedicated for WSN applications are available [42] and can have 4 to 32 data bits, 512 Bytes to 128 KB RAM, 1 to 6 timers, power dissipation as low as 1.9 mW and the appropriate choice can be made based on the requirements.

9.5.2 FPGA

The FPGA technology provides the ability to develop customized WSN nodes and a dedicated single-purpose processor can be developed in HDL along with the required timers, counters, and communication protocols such as UART and USB. Moreover, the soft-core of an already developed general purpose or an application processor can also be tweaked to meet the design requirements. The major challenge in this regard is the comparatively high power dissipation. For this purpose, ultra-low-power FPGA units need to be used for WSN node development. The main benefit of using FPGA compared to microcontrollers for implementing WSN nodes is their performance as has been demonstrated in Hamila et al. [43]. The high performance feature of FPGAs can be leveraged upon to build on-chip image processing sensors [44] for WSN nodes.

9.5.3 WSN Prototype Implementation

In general, we recommend using Flash-based FPGA, such as ACTEL Fusion FPGA module, for developing the desired SNs in WSNs designed for PP. The availability of Flash would allow using low power sleep modes instead of the wake-up reconfiguration that is required in the other types of FPGAs. We also recommend using the ARM Cortex M1 core on our FPGA and this decision is mainly motivated by the high performance and small size of this core, which is desirable for PP applications. This processor also provides high and a low speed buses. For the analog components, like ADC and multiplexer, numerous off-the-shelf components can be placed on the same board as the FPGA. The required filters can be implemented on the FPGA. These FPGA-based SNs will be remotely accessed by researchers and/or public users via the internet connection as shown in Figure 9.7. Moreover, advances in cloud computing will empower such kind of light-weight networks, i.e., the WSNs, by offloading the heavy analysis and processing part up to the High Performance Computing (HPC) machines to filter and process raw data coming directly from the field. Accordingly, farmers and researchers can collect comprehensive and precise yield data without significant efforts and inexpensively as shown in Figure 9.8. Yield maps can be generated in real-time subsequent to data collection to identify yield general patterns within fields. These maps allow recognizing within-field spatial variability for variable rate applications, empowering farmers to estimate the economic revenues of different farming management plans. Furthermore, they are essential for field-level developments such as leveling the land, timing of irrigation systems, drainage, building fences, and for off the field data usage.

The information provided in PP represents a valuable resource because it enables real-time decision making with regard to critical issues, such as establishing water saving policies while providing adequate irrigation, and choosing the right time for farming activities such as planting, harvesting, specifying the fruit maturity etc. In order to provide services for relatively large areas of coverage, it may be necessary to employ large numbers of WSN nodes, which use applications and communication protocols tailored mainly to provide higher energy efficiency. A popular approach in energy efficient routing is to use cooperative game theory. Recently, this theory is widely used in studies associated with wireless networks [45]. This is because energy is highly wasted in such networks due to extensive cooperation in data forwarding. Therefore, the game theoretic mechanism which applies the concept of gains and losses has been implemented in AlSkaif et al. [46] for WSNs. Game theory can also be used to analyze the gains a WSN node can make and can certainly lead toward the achievement of equilibria for all the concerned nodes. Moreover, game theory has been used in various articles for selfish nodes management. Incentive-based mechanisms have been successfully applied in many approaches for load balancing, energy efficiency, etc., in WSNs. Such mechanisms are categorized into credit-based and reputation-based [47]. The reputation-based mechanism relies on the

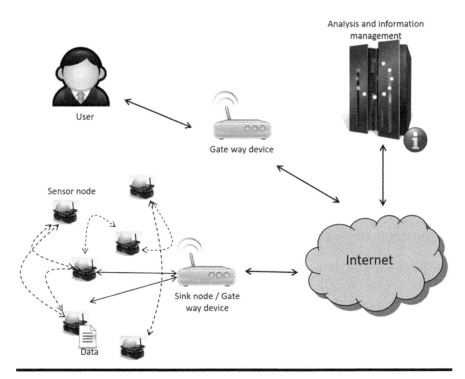

Figure 9.8   WSN architecture for PP applications.

evaluation of nodes' behavior. In this type of approach, different reputation stages are made to determine the nodes' cooperation level [48]. The message forwarding by the WSN intermediate nodes are made according to a reputation value at the source [49, 50]. In the credit-based approach, a WSN node can gain credit-scores by offering these relaying services (Figure 9.9).

It's worth pointing out that the initial deployment and routing based decisions for the WSN nodes and the configuration of wireless communications related components can make significant differences for the evaluation process in PP. Starting the evaluation process by employing the prototype straight from the beginning and checking various strategies can be quite costly especially in terms of time and efforts. Therefore, similar to the existing studies in the literature, we believe that it can be ideal to use simulation tools prior to deployment in an attempt to have a degree of optimization beforehand.

9.6 Concluding Remarks

This article aims at initiating further research in many directions related to WSNs applications in PP. These directions include the MAC layer design, security, energy efficiency, cross-layer optimization, routing and scheduling, performance evaluation, and complexity reduction. Other related research areas include sensors' design, data

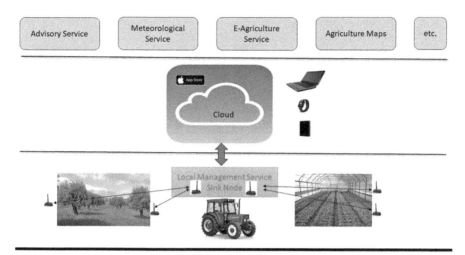

Figure 9.9   A block diagram of the WSN nodes in the Cloud.

processing, cloud computing, data fusion, positioning and localization, and formal performance analysis.. It is necessary to emphasize that PP is currently receiving increasing attention by the governmental and industrial sectors in several countries in the world including the US, Canada, Brazil, Malaysia, India, and many other countries. In such countries, a substantial effort and focus is directed toward the research and integration of PP technologies. The same focus and attention to PP is also given by several industrial leaders such as IBM [51], AG Leader [52], and Precision Planting [53]. The immediate next steps toward involving WSNs in large scale PP deployments, is the incorporation of the state-of-the-art technology such as IoT and Cloud computing.

As a future work, we foresee the immediate next steps to include field studies with large scale deployments, data collection and mining, in addition to incorporating the state-of-the-art information technology such as the Internet of Things and cloud/edge computing in order to further improve the performance. These kinds of technologies are expected to have a significant impact on the field of PP in particular, and wireless networks in general. Therefore, the output of any research in this direction will initiate further advances in many directions related to the PP applications. Furthermore, the measurement criteria for the success of any PP underlying infrastructure will be based on the energy efficiency of the custom nodes, and MAC/Routing algorithms, as well as other well-known QoS related measures which are relevant to agriculture applications such as delay, consistency, and reliability.

References

1. F. Yalçınkaya, N. Çakar, M. Mısırlıoğlu, N. Tümer, N. Akar, M. Tekin, H. Taştan, H. Koçak, N. Özkaya, A. H. Elhan, "Genotype–Phenotype Correlation in a Large Group of Turkish Patients with Familial Mediterranean Fever: Evidence for Mutation-Independent Amyloidosis," *Rheumatology*, vol. 39, no. 1 (2000): 67–72.

2. F. Al-Turjman, "Modelling Green Femtocells in Smart-grids," *Springer Mobile Networks and Applications*, vol. 23, no. 4 pp. 940–955, 2018.
3. F. Al-Turjman, "Mobile Couriers' Selection for the Smart-grid in Smart cities' Pervasive Sensing," *Elsevier Future Generation Computer Systems*, vol. 82, no. 1 pp. 327–341, 2018.
4. F. Al-Turjman, "Energy -aware Data Delivery Framework for Safety-Oriented Mobile IoT," *IEEE Sensors Journal*, vol. 18, no. 1 pp. 470–478, 2017.
5. S. Alabady and F. Al-Turjman "LCPC Error Correction Code for Internet of Things Applications", *Elsevier Sustainable Cities and Society*, vol. 42, pp. 663–673, 2018.
6. M. Biglarbegian and F. Al-Turjman "Path Planning for Data Collectors in Precision Agriculture WSNs", *In Proc. of the International Wireless Communications and Mobile Computing Conference (IWCMC)*, Nicosia, Cyprus, 2014, pp. 483–487.
7. F. Al-Turjman "Cognitive Caching for the Future Sensors in Fog Networking", *Elsevier Pervasive and Mobile Computing*, vol. 42, pp. 317–334, 2017.
8. R. Heuvel, "The Promise of Precision Agriculture," *Journal of Soil and Water Conservation*, vol. 51, no. 1 (1996): 38.
9. D. Berckmans, "Automatic On-line Monitoring of Animals by Precision Livestock," *Proceedings of the International Society for Animal Hygiene*, Saint-Malo, (2004): 27–30.
10. Mutka and R. Bart, "Image-based Phenotyping of Plant Disease Symptoms,", Front Plant Sci.2014; vol. 5, no. 1, pp. 734–745.
11. J. Humplik, D. Lazar, A. Husickova and L. Spichal,"Automated Phenotyping of Plant Shoots using Imaging Methods for Analysis of Plant Stress Responses—a Review,", *Plant Methods*, vol. 11, no. 29 (April, 2015): 1–10.
12. J.Araus and J. Cairn,"Field High-throughput Phenotyping: The New Crop Breeding Frontier," *Trends in Plant Science*, vol 19, no. 1 (January, 2014): 52–61.
13. L. Li, Q. Zhang, and D. Huang, "A Review of Imaging Techniques for Plant Phenotyping," *Sensors*, vol. 14, no. 11, pp. 20078-20111 (2014): 20078–20111.
14. F. Al-Turjman and H. Hassanein, "Towards Augmented Connectivity with Delay Constraints in WSN Federation," *International Journal of Ad Hoc and Ubiquitous Computing*, vol. 11, no. 2/3 (2012): 97–108.
15. F. Al-Turjman, A. Alfagih, W. Alsalih, and H. Hassanein, "Reciprocal Public Sensing for Integrated RFID-Sensor Networks," *Proceedings of the IEEE International Wireless Communications & Mobile Computing Conference (IWCMC)*, Cagliari, Sardinia, Italy, (2013): 746–751.
16. M. Ishizuka and M. Aida, "Performance Study of Node Placement in Sensor Networks," *Proceedings of 24th International Conference on Distributed Computing Systems Workshops (Icdcsw)*, vol. 7, March, 2004 Tokyo, Japan.
17. M. Younis and K. Akkaya, "Strategies and Techniques for Node Placement in Wireless Sensor Networks: A Survey," *Elsevier Ad Hoc Network Journal*, vol. 6, no. 4 (2008): 621–655.
18. F. Al-Turjman, H. Hassanein, and M. Ibnkahla, "Efficient Deployment of Wireless Sensor Networks Targeting Environment Monitoring Applications: A Survey," *Elsevier Ad Hoc Network Journal*, vol. 6, no. 4 (2008): 621–655.
19. M. Jongerden and B.R. Haverkort, "*Battery Modeling*," Technical Report TR-CTIT-08-01, 2008.
20. J. Aldous and R. Wilson, *Graphs and Applications: An Introductory Approach*. Verlag: Springer, 2000.

21. Y. Chang and C. Hsu, "Routing in Wireless/Mobile ad-hoc Networks via Dynamic Group Construction," *Mobile Networks and Applications*, vol. 5 (2000): 27–37.
22. A. Abdrabou and W. Zhuang, "A Position-based QoS Routing Scheme for UWB ad hoc Networks," *IEEE Journal on Selected Areas of Communications*, vol. 24, no. 4, April 2006.
23. N. Ray and A. Turuk, "A Review on Energy Efficient MAC Protocols for Wireless LANs," *Proceedings of 4th International Conference Industrial and Information Systems*, (December, 2009): 137–142 Sri Lanka, Sri Lanka.
24. S. Choudhury and F. Al-Turjman, "Dominating Set algorithms for Wireless Sensor Networks Survivability" *IEEE Access Journal*, vol. 6, no. 1 (2018): 17527–17532.
25. J. Zhang, G. Zhou, C. Huang, S. H. Son, J. A. Stankovic, "TMMAC: An Energy Efficient Multi-Channel MAC Protocol," *Proc. IEEE ICC*, 2007 Glasgow, UK.
26. F. Al-Turjman, "QoS–aware Data Delivery Framework for Safety-inspired Multimedia in Integrated Vehicular-IoT," *Elsevier Computer Communications Journal*, vol. 121 (2018): 33–43.
27. J. Smith, "M/G/c/K Blocking Probability Models and System Performance," *Performance Evaluation*, vol. 52 (2003): 237–267.
28. A. Singh, H. Tiwari, A. Vajpayee, and S. Prakash, "A Survey of Energy Efficient Routing Protocols for Mobile ad-hoc Networks," *International Journa of. Computer Science & Engineering*, vol. 2, no. 99 (2010): 3111–3119.
29. Y. Wang, X. Li, W. Song, M. Huang, and T. Dahlberg, "Energy-efficient Localized Routing in Random Multihop Wireless Networks," *IEEE Transactions on Parallel and Distributed Systems*, vol. 22, no. 8 (August, 2001): 1249–1257.
30. M. Z. Hasan and F. Al-Turjman, "SWARM-based Data Delivery in Social Internet of Things," *Elsevier Future Generation Computer Systems*, 2017. Doi:10.1016/j.future.2017.10.032.
31. Y. Hu, Y. Ding, and K. Hao, "An Immune Cooperative Particle Swarm Optimization Algorithm for Fault-tolerant Routing Optimization in Heterogeneous Wireless Sensor Networks," *Journal Mathematical Problems Engineering*, vol. 2012 (August, 2011): 19.
32. M. Dehghan, M. Ghaderi, and D. Goeckel, "Minimum-energy Cooperative Routing in Wireless Networks with Channel Variations," *IEEE Transactions on Wireless Communications*, vol. 10, no. 11 (Novemer, 2011): 3813–3823.
33. Y. Kim, E. Lee, and H. Park, "Ant Colony Optimization Based Energy Saving Routing for Energy-efficient Networks," IEEE Communications Letter, vol. 15, no. 7 (July, 2011): 779–781.
34. C. Ma and Y. Yang, "A Battery-aware Scheme for Routing in Wireless ad hoc Networks," *IEEE Transactions on Vehicular Technology*, vol. 60, no. 8 (October, 2011): 3919–3932.
35. F. Ren, J. Zhang, T. He, C. Lin, and S. K. Das, "EBRP: Energy-balanced Routing Protocol for Data Gathering in Wireless Sensor Networks," *IEEE Transactions on Parallel and Distributed Systems*, vol. 22, no. 12 (December, 2011): 2108–2125.
36. S. Alabady and F. Al-Turjman, "Low Complexity Parity Check Code for Futuristic Wireless Networks Applications", *IEEE Access Journal*, vol. 6, no. 1, pp. 18398–18407, 2018.
37. M. Z. Hasan, and F. Al-Turjman, "Analysis of Cross-layer Design of Quality-of-service Forward Geographic Wireless Sensor Network Routing Strategies in Green Internet of Things," *IEEE Access Journal*, vol. 6, no. 1 (2018): 20371–20389.

38. D. Chen, H. Ji, and X. Li, "An Energy-efficient Distributed Relay Selection and Power Allocation Optimization Scheme over Wireless Cooperative Networks," *IEEE ICC*, (June, 2011): 1–5 Kyoto, Japan.

39. F. Al-Turjman, and S. Alturjman, "5G/IoT-enabled UAVs for Multimedia Delivery in Industry-oriented Applications," *Springer's Multimedia Tools and Applications Journal*, 2018.

40. M. Ismail and W. Zhuang, "A Distributed Multi-service Resource Allocation Algorithm in Heterogeneous Wireless Access Medium," *IEEE Journal on Selected Areas of Communications*, vol. 30, no. 2 (February, 2012): 425–432.

41. F. Al-Turjman, "Positioning in the Internet of Things Era: An Overview," In *Proc. of the IEEE International Conference on Engineering and Technology (ICET)*, Antalya, Turkey, 2017.

42. M. Lakhzouri, Simona Lohan, Ridha Hamila, and Markku Renfors, "Extended Kalman Filter for LOS Estimation in WCDMA Mobile Positioning," in *EURASIP Journal on Applied Signal Processing*, vol. 13 (December, 2013): 1268–1278.

43. U. Ulusar, G. Celik, F. Al-Turjman, "Wireless Communication Aspects in the Internet of Things: An Overview", In *Proc. of the IEEE Local Computer Networks (LCN)*, Singapore, 2017.

44. S. Demir and F. Al-Turjman, "Energy Scavenging Methods for WBAN Applications: A Review", *IEEE Sensors Journal*, vol. 18, no. 16, pp. 6477-6488, 2018.

45. L. Dasilva, H. Bogucka, and A. Mackenzie, Game Theory in Wireless Networks, *IEEE Communications Magazine*, vol. 49, no. 8 (August, 2011): 110–111.

46. T. AlSkaif, M. Guerrero Zapata, and B. Bellalta, Game Theory for Energy Efficiency in Wireless Sensor Networks: Latest Trends, *Journal of Network and Computer Applications*, vol. 54 (August, 2015): 33–61.

47. Z. Chu1, H. X. Nguyen1, T. A. Le1, M. Karamanoglu1, D. To, E. Ever, F. Al-Turjman and A. Yazici, "Game Theory Based Secure Wireless Powered D2D Communications with Cooperative Jamming", *IEEE Wireless Days conference,* Porto, Portugal, 2017, pp. 95–98.

48. M. Umar, S. Khan, R. Ahmad, and D. Singh, "Game Theoretic Reward Based Adaptive Data Communication in Wireless Sensor Networks," *IEEE Access*, 2018. Doi:10.1109/ACCESS.2018.2833468

49. D. Yang, X. Fang, and G. Xue, Game Theory in Cooperative Communications, *IEEE Wireless Communications*, vol. 19, no. 2 (April, 2012): 44–49.

50. F. Al-Turjman, "Fog-based Caching in Software-Defined Information-Centric Networks", *Elsevier Computers & Electrical Engineering Journal*, vol. 69, no. 1, pp. 54–67, 2018.

51. F. Al-Turjman, "Price-based Data Delivery Framework for Dynamic and Pervasive IoT", *Elsevier Pervasive and Mobile Computing Journal*, vol. 42, pp. 299–316, 2017.

52. F. Al-Turjman, "Optimized Hexagon-based Deployment for Large-Scale Ubiquitous Sensor Networks", *Springer's Journal of Network and Systems Management*, vol. 26, no. 2, pp. 255–283, 2018.

53. F. Al-Turjman, "Information-Centric Framework for the Internet of Things (IoT): Traffic Modelling & Optimization", *Elsevier Future Generation Computer Systems*, vol. 80, no. 1, pp. 63–75, 2017.

Chapter 10

Intelligent Positioning for Precision Agriculture (PA) in Smart-cities

Fadi Al-Turjman and Sinem Alturjman*

Contents

* Antalya Bilim University, Antalya, Turkey

10.1 Introduction

Positioning systems used in Wireless Sensor Networks (WSNs) deployed for gathering data in smart-cities' Precision Agriculture (PA), viz. smart farming and crop harvesting is a challenging problem where wireless nodes equipped with sensors and GPS modules are subject to several risks. Firstly, these nodes may be covered and lose the line of sight with the satellite by extreme weather conditions such as rain and snow. Secondly, the wireless communication channels are highly affected by the growing plants and areas of dense trees. This makes data source positioning one of the most challenging issues to be considered in the PA applications.

Global Positioning System (GPS) provides positioning, velocity, and time information with consistent and acceptable accuracy when there is direct line of sight to four or more satellites [1, 2]. However, GPS may suffer from outages, jamming, and multipath effects in urban areas, canyons, and rural foliage canopies as in PA applications. Inertial Navigation System (INS), on the other hand, is self-contained meaning it is immune to external interference, but its accuracy deteriorates in the long term due to sensor's bias, drift, scale factor instability, and misalignment [3–5]. By integrating the GPS and INS signals, a complementary solution can be obtained that is often more accurate than that of each independent system [3]. Kalman Filter (KF) is traditionally used to optimally fuse the position and velocity information from both INS and GPS [6–9]. However, cost and space constraints are the two primary obstacles that have prevented the utilization of either navigation or tactical grade INS inside the node in WSNs utilized in PA applications. There are several inadequacies in the KF technique. Such inadequacies include: (1) Observability problems, (2) Error modeling challenges, and (3) Poor prediction during GPS outages. In order to avoid these problems, and to provide an integrated system that can be independent from the underlying navigation system, and design, an Artificial Intelligent (AI) solution has been used to solve this problem. In general, the main objective of AI is to learn the pattern of the data and build the path accordingly. The objective is to use this knowledge, when GPS signal is not available, to predict the trajectory path. The network model of the multilayer perceptron architecture is based on Neural Network (NN) units which compute a nonlinear function of the scalar product of the input vector and the weight vector. An alternative architecture of a NN is one in which the distance between the input vector and a certain prototype vector, determines the activation of a hidden NN unit.

In this chapter, we propose an enhanced Radial Basis Neural Network Function (RBNNF)-based positioning system to integrate INS and GPS. The presented system will enable more efficient positioning in agriculture applications in a real-time manner. The system involves training the network in real time, when the GPS signal is available, and predicting proper positioning, when GPS signal is not available.

The remainder of this chapter is organized as follows. In the next section, a literature review is presented. Section 10.3 describes system models used to implement the RBNNF system. In Section 10.4, we discuss the methodology followed in this system. Experimental results and discussions are detailed in Section 10.5. Finally, conclusions are provided in Section 10.6.

10.2 Literature Review

Recently work has been done to achieve INS/GPS integration using NN. Multilayer perceptron (MLP) NNs have been presented in Chiang et al. and Noureldin et al. [4, 5]. The authors were able to show that a Position and Velocity Update Architecture (PVUA) could provide the position components along both the east and north directions [4]. However, their approach provided only the INS position components, but not the errors in predicting these position components [6]. Therefore, accuracy of this system could not be measured. Moreover, no real-time implementation has been addressed in Chiang et al. [6].

The next attempt was done in Noureldin et al. [5], where the authors proposed a P_P model. The proposed model provides the INS position error at the output [5]. The suggested method feeds three P_P networks (one for each position component) by GPS position updates. The problem with this approach is that internal structure of each of the MLP networks has to be changed until reaching the best performance, which leads to long training time [6]. Moreover, the authors didn't address the issue of the real-time implementation [6].

To address the issue of long training time, and real-time processing, NNs has been proposed instead of MLP networks [6]. NN is utilized due to its massively parallel-distributed processors that can be used to model highly complex and non-linear stochastic problems that cannot be solved using conventional algorithms [7–9].

The major benefit of RBFNN is its ability to utilize Radial Basis Function (RBF) networks without identifying the number of neurons in its hidden layer [6]. The idea is that hidden neurons are dynamically generated during the training procedure to achieve the desired performance. The major problem in this approach is the long time needed for training the RBF, which was impractical and almost impossible to implement in real time [6]. Moreover, the proposed system did not address real-time implementation, and factors that affect the performance of the system in real time [6].

In order to address real-time implementation issues, an Adaptive Real-Time Neuro Fuzzy Inference System (ANFIS)-based system has been proposed for mobile multi sensor INS/GPS navigation system integration [6]. This approach, however, suffered huge computation load for real-time implementation, causing limitations regarding the optimization of the ANFIS parameters during real-time operations.

Finally, to resolve the issue of real-time implementation, an RBFNN-based module for real-time INS/GPS integration was proposed [6]. The major benefits of RBFNN are simple implementation, fast training, and the capability to achieve high levels of accuracy [7–9]. Moreover, there is a need only to modify a small number of parameters inside the RBFNN module, which means, low computational load, and much better real-time performance. However, the methodology suffers from increase in error as time goes while GPS signal is not available.

In this chapter, we propose an enhanced RBFNN to overcome the problems in the literature. INS and GPS position data is fused together, to predict the INS position error. Moreover, the proposed system will predict the position during GPS outages, based on processing only INS position components. We implement and simulate the above proposed system, while not including velocity data. This is done in order to avoid further computational load in real-time mode. Moreover, the modified algorithm for simulating GPS outage and predicting the correct position has been modified to reduce the increase in prediction error over time.

10.3 Enhanced Radial Basis Neural Network Function (RBNNF)-Based Approach

The proposed system architecture compromises two modes of operation: The training mode, and the prediction mode. The update mode is activated as long as the GPS signal is available. During this mode of operation, the RBFNN is trained and the optimal values of its internal parameters are determined.

The INS position is the input to the module while the error in the INS position is the module output. At each time instant, the estimated INS position error provided by the RBFNN module is then compared to the error between the INS original position and the corresponding GPS position. The difference is the estimation error of the RBFNN module. In order to minimize this error, the training of the RBFNN module continues and the RBFNN parameters are continuously updated according to the least-square criterion until reaching certain minimal mean-squared estimation error.

NN has the ability to continuously adapt its internal structure to the changing node dynamics. In this chapter, simple two-layer RBFNN architecture is utilized. The first layer (known as hidden layer) contains a set of basic functions, and the second layer (known as output layer) forms the linear combinations of these basis functions to generate the output [7–9]. This unique architecture of the RBFNN has the advantage of a fast and real-time training procedure.

Upon losing the satellite signal (during GPS outages), the system switches automatically to the prediction mode where the RBFNN module is used to predict the INS position error using the latest RBFNN parameters obtained before losing the satellite signals. The error is then removed from the corresponding INS position component to obtain the corrected INS position. In order to provide a complete navigation solution in the two axes for a moving node, two modules are utilized for the two position components defined along the north and east directions. Note that vertical direction could be easily added in another implementation.

In order to simulate the GPS outage and be able to predict it properly, a NN should be created for every dimension. Then the NN is trained using the available data, and based on proper window size (as discussed above). GPS signal is simulated to be 0, then NN is used to predict the error in predicting INS signal. The reason behind this is that the predicted error is used to calculate proper positioning. Finally, the path is plotted using Google Maps. The algorithm is summarized in Algorithm 1.

In line 8 of the algorithm, RBFNN is trained in supervised learning, with the parameters of the Gaussian basis functions being set as default by MATLAB. The Gaussian functions are used as the receptive units inside the hidden layer. They act as the operator providing the information about the class to which the input INS position signal belongs. Each of the Gaussian RBFs is defined by its center and spread set as defaults by MATLAB.

ALGORITHM 1: SIMULATE GPS OUTAGE AND PREDICT INS.

1. Initialize earth eccentric
2. Initialize window size
3. Repeat for every epoch:
4.   Fill the window of NN learning in:
5.   Store INS and GPS 3-dimensional data
6.   Store absolute (difference in measurement/2)
7.   If window is filled, add items at top window (FIFO)
8.   If epoch is designated outage simulation epoch do:
9.   Create RBFNN for every dimension
10.   For the next 30 epochs do:
11.   Predict error in INS position
12.   Calculate corrected position
13.   Store in temp storage
14. Convert measures to meters
15. Plot the Graphs using Temp Storage

In line 9 above, RBFNN module for INS/GPS integration takes the INS position at the input and provides the corresponding position error at the output. In line 4 above, the proper choice of the window size is essential to guarantee delivery of the desired accuracy while ensuring system robustness in real time. There is a trade-off in choosing small or large window sizes. For instance, large window size is beneficial in capturing the most recent node dynamics, thus the RBFNN module becomes more reliable during long GPS outages.

Large window size may complicate the update procedure and result in long training time, which is not suitable for real-time implementation. On the other hand, the reduced level of the non-stationary nature of INS and GPS data in case of small window size makes the RBFNN update procedure faster and more robust for real-time operation. However, small window size may cause the system to become less reliable in case of relatively long GPS outages exceeding the double of its size.

In line 6, the difference in measurement (between INS and GPS) is further divided by 2 and the absolute value of the result is stored and used in line 11 to predict the error in INS position. This modification of the algorithm is designed to reduce the difference in error. This error is calculated at line 6 above.

However, this causes the error to increase over time [6]. In the modified algorithm, this error is divided and the absolute error is stored. This modification caused slowdown of the increase in prediction error over time. The enhancement is exhibited by showing the degree at which the error is stable, before it starts to increase. More stable error has been reached, and less adrift INS error prediction has been obtained.

10.4 Experimental Results and Discussion

The performance of the enhanced RBFNN-based module was examined using three simulated GPS outages. These GPS outages were intentionally introduced for both trajectories. The locations of these GPS outages were chosen to examine the performance of the RBFNN module with respect to different node dynamics. For the purpose of real-time implementation, a window size of 30 seconds was chosen for two outages, and 50 seconds for one outage.

In these three planned outages, the reference position will be the same trajectory when the GPS signal is available. Using this reference, we can find the quality and level of accuracy of the predicted trajectory.

10.4.1 Setup and Simulation Environment

Using MATLAB, we simulate GPS outage by assigning 0 to the GPS data signal, and simulate prediction of GPS signal using MATLAB NEWRB Function. The simulation is based on predicting the error between GPS signal and INS signal. This predicted error is then used to calculate GPS signal, hence arriving at the proper positioning.

The learning window size is assigned as discussed in the previous section. The presented system is simulated and tested using real measurements from an inertial sensor and GPS mounted on a land vehicle. To evaluate the predicted path, the original plot (where GPS signal is available) is plotted against the predicted path, for the same trajectory. All plots were done using Google Earth.

10.4.2 Metrics and Parameters

Gaussian functions in this simulation are all assumed to be 0 cantered with 1 degree of Standard Deviation. This is the default definition of the NEWRB function in MATLAB. These defaults worked fine for the purpose of this simulation.

A next stage development would be to learn these parameters in an unsupervised learning. The input for NEWRB is the Longitude and Latitude values stored in the learning window, whereas the output is the predicted Error in predicting Latitude and Longitude. The Windows size is assigned to be 30 in two cases and 50 in one case. This is the best size that captures the dynamics of GPS signals, while not being overloaded in computation.

10.4.3 Simulation Cases

GPS first outage was planned and simulated at second# 152540. The outage duration was 30 seconds (1 epoch/second). The training window size was 30 seconds (1 epoch/second). Altitude was ignored (assumed to be 0). The Reference points were actual data from the GPS for the same period of outage simulation. The second outage was at second: 153641. It was equal to 30 seconds. The training window was equal to 50 seconds. The third outage was at second: 154541. It was equal to 30 seconds and the training window was 50 seconds. Next, the trajectory path was presented for both, when the GPS signal and the predicted path was available. Two plots are presented for each case.

Trajectory (Figure 10.1) exhibits a large degree of accuracy, smoother than GPS (Figure 10.2) path. Average Longitude Prediction Error was 8.7 m, Std. Dev is 8.7 m, exhibiting good accuracy. This could be attributed to capturing the most recent GPS dynamic by the learning NN. For Latitude the Average Error is 130.7 m, Std. Dev. 102.8 m. The graph of Longitude errors (Figure 10.3) exhibits consistent increasing error, reflecting the most recent dynamic of the GPS as trained by the NN. The Latitude error graph (Figure 10.4) shows no stable pattern, reflecting the non stable dynamics of the most recent GPS signal.

Prediction (Figure 10.5) shows a large degree of inaccuracy compared to (Figure 10.6). This could be attributed to a changing of GPS path dynamics.

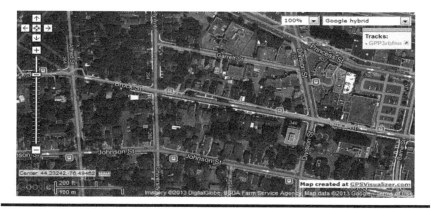

Figure 10.1   Predicted RBFNN Trajectory—Case 1.

Figure 10.2   GPS Trajectory—Case1.

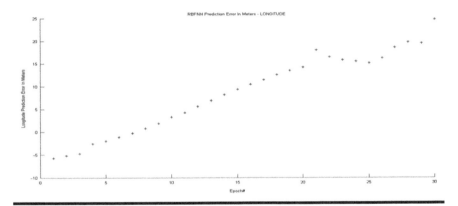

Figure 10.3   Prediction Error Plot—Longitude vs Epoch# Case1.

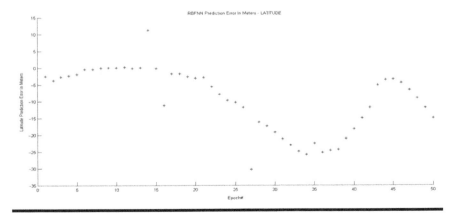

Figure 10.4   Prediction Error Plot—Latitude vs Epoch# Case1.

Figure 10.5 Predicted Trajectory—Case 2.

Figure 10.6 GPS Trajectory—Case 2.

The Average Latitude error is 14.9 m whereas the Std. Dev. is 9.5 m. The graph (Figure 10.7) shows stability first, then steep upward movement, reflecting the dynamic of recent GPS as captured by the learning NN. The average Longitude is 123.4 m, and Std. Dev. is 147.4 m, thus causing the inaccuracy of the path. Graph (Figure 10.8) show stable upward errors, also due to the dynamic of recent GPS signal captured by the learning NN.

Trajectory (Figure 10.9) shows a large degree of accuracy, better than GPS path (Figure 10.10). Average Lon Error = 19.5 m, Std. Dev. = 28.6m. This exhibits good level accuracy, reflecting similar dynamic of recent GPS signal trained by the NN. For Latitude, it's 105.4 m, Std. Dev. 122.7 m. Not as good results. Error graph (Figure 10.11) shows good stability at first, then modest and slow increase, reflecting the similar dynamic of recent GPS signal trained by the NN. For Latitude it's 105.4 m, Std. Dev. 122.7 m. The graph (Figure 10.12) shows steep upward increase, due to capturing the most recent dynamic of GPS.

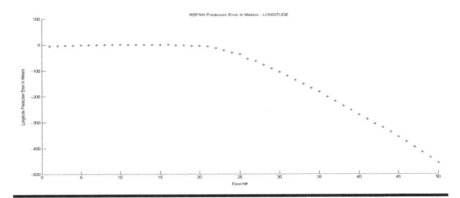

Figure 10.7 **Prediction Error Plot—Longitude vs. Epoch# Case2.**

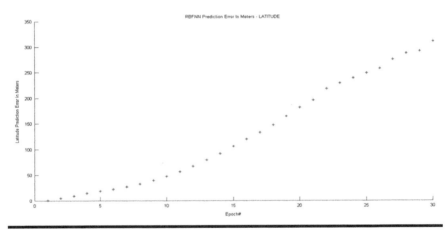

Figure 10.8 **Prediction Error Plot—Latitude vs Epoch# Case2.**

Figure 10.9 **Predicted Trajectory—Case3.**

Figure 10.10   GPS Trajectory—Case3.

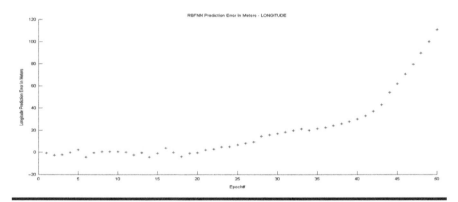

Figure 10.11   Prediction Error Plot—Longitude vs Epoch# Case3.

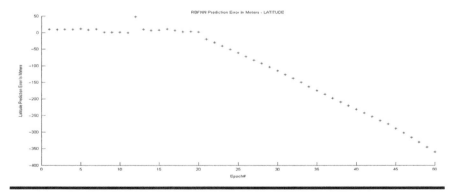

Figure 10.12   Prediction Error Plot—Latitude vs Epoch# Case3.

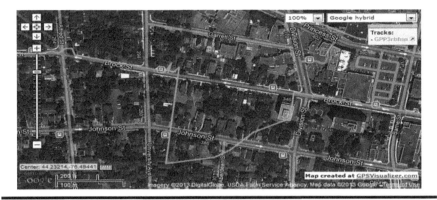

Figure 10.13 Predicted Trajectory—window size 50 s.

Figure 10.14 Predicted Trajectory—window size 30 s.

Also, the size of training window is critical for proper training of RBFNN and therefore, best prediction of INS position errors. For example, the first case (Figure 10.13) shows great drift away from path, due to a change in dynamics of GPS. While in (Figure 10.14) much better results were obtained due to capturing dynamics in a limited range.

10.5 Conclusions

In this chapter, the real-time INS and GPS integration in navigation utilizing RBFNN, has been simulated and implemented. This architecture is based on predicting the INS position error and continuously removing it from its corresponding INS position. The simulation was employed in real time using MATLAB.

Results showed the ability of the RBFNN-based module to reduce the INS position error and prevent its growth even in the long term. We saw in the cases that error is stabilized and takes a pattern in time. In addition, the proposed RBFNN module was able to accurately predict the INS position errors during GPS outages. This was the objective of the system as introduced early in this chapter. Results point to success in achieving these objectives, of being able to predict, in real time, the proper position of a vehicle, based on learning the dynamics of the most recent GPS signal, with a good degree of accuracy.

The simulation proved that the integration technique is a reliable, robust, and self-adaptive system that requires no prior knowledge of the navigation system utilized. A Future research could focus on what are reasons where dynamics of GPS caused that error in the second case, above predicting the path. Moreover, it could also investigate the 3rd dimension (altitude) which was not covered in this chapter. Another area of improvement in future research, is further reducing the increase in prediction error over time.

References

1. M. S. Grewal, L. R. Weill, and A. P. Andrews, *Global Positioning Systems, Inertial Navigation and Integration*. New York: Wiley, 2001.
2. D. H. Titterton and J. L. Weston, *Strapdown Inertial Navigation Technology*. London, UK: Peregrinus, 1997.
3. A. Noureldin, D. Irvine-Halliday, and M. P. Mintchev, "Accuracy Limitations of FOG-based Continuous Measurement while Drilling Surveying Instruments for Horizontal Wells," *IEEE Transactions on Instrumentation and Measurement*, vol. 51, no. 6 (December, 2002): 1177–1191.
4. K. W. Chiang, A. Noureldin, and N. El-Sheimy, "Multi-sensors Integration using Neuron Computing for Land Vehicle Navigation," in *GPS Solutions*, Heidelberg Germany: Springer-Verlag, vol. 6 (March, 2003): 209–218.
5. A. Noureldin, A. Osman, and N. El-Sheimy, "A Neuro-wavelet Method for Multi-sensor System Integration for Vehicular Navigation," *Journal of Measurement and Science Technology*, vol. 15, no. 2 (February, 2004): 404–412.
6. K. W. Chiang, A. Noureldin, and N. El-Sheimy, "The Utilization of Artificial Neural Networks for Multi-sensor System Integration in Navigation and Positioning Instruments"; *IEEE Transactions on Instrumentation and Measurement*, April 2003.
7. R. Sharaf, A. Noureldin, A. Osman, N. El-Sheimy, "Online INS/GPS Integration with a Radial Basis Function Neural Network," *IEEE Aerospace Electronic and Systems Magazine*, vol. 20, no. 3 (March, 2005): 8–14.
8. C. M. Bishop, *Neural Networks for Pattern Recognition*. London, UK: Oxford Univ. Press, 1995.
9. F. Al-Turjman "QoS -aware Data Delivery Framework for Safety-inspired Multimedia in Integrated Vehicular-IoT," *Elsevier Computer Communications Journal*, vol. 121, pp. 33–43, 2018.
10. F. Al-Turjman, H. Hassanein, S. Oteafy, and W. Alsalih, "Towards Augmenting Federated Wireless Sensor Networks in Forestry Applications," *Personal and Ubiquitous Computing*, vol. 17, no. 5 (June, 2013): 1025–1034.

Chapter 11

Security Issues in VANET-based Smart-cities

Fadi Al-Turjman*, Cem Serimözü†,
and Chadi Altrjman‡

Contents

* Antalya Bilim University, Antalya, Turkey
† Middle East Technical University, Ankara, Turkey
‡ University of Waterloo, ON, Canada

11.1 Introduction

Considering the rate at which VANETs have been developing and the benefits expected from vehicular interconnections with millions of vehicles in traffic worldwide, it is most likely that they will be the most concrete and common usage of mobile ad-hoc networks (MANETs) [1, 2]. With the future possibility of seamless connectivity expected from 5G with high backward compatibility and highly integrative design [3], proper implementation and dissemination of Onboard Units (OBUs), geolocational devices such as GPS or GLONASS (Global Navigation Satellite System) receivers, combined with the ever-increasing communicational potential, create significant opportunities be they economic or technological, but also raises interesting research challenges and important ethical questions [4, 5].

The increasing connectivity between numerous elements has made efficient security protocols and applications a prime subject of study. With the massive amount of data that can be produced by modern vehicles that should be treated in a timely manner, efficiency is without a doubt paramount in a system where an error or an attack can result in the loss of human life. One of the primary constraints of VANETs compared to most systems is making sure life-critical information is delivered on time while establishing the liability of the users and simultaneously protecting the privacy of the drivers to its fullest extent. Balancing both the constraints of this medium with effective security requires us to fulfill numerous parameters without exceeding the need of one and vice versa. VANETs by their nature have some advantages and disadvantages compared to most MANETs and IoT networks. Detailed analysis of the properties of VANETs needs to be established to develop flexible and modular solutions [6].

The type of communication transmitted in the network has different constraints and urgency in accordance with the content of the communicated information. The way security is applied needs to be in proportion to the confidentiality required for that specific message. This creates a need to classify different levels of communication before even broadcasting them. Proper privacy enforcement requires cooperation between the governmental institutions and private companies. The ethical and technical applications, both theoretical and practical, needed for effective treatment of information while conserving enough privacy needs in-depth discussions. Understanding the nature of potential attacks and disturbances both intentional and accidental has made it a requirement for researchers on the topic to classify them according to different criteria.

In the first section, we will detail our system model by defining what are the characteristics, both positive and negative, of VANET. We will also discuss the message constraints and the message types that will be delivered across the network to identify critical and constrained messages. The second section contains a general attacker classification, attacks discussed in the literature, and potential defense methods that have been applied in particular to them. The third part contains the overall classification of defense methods and the presentation of some of the techniques that fall under that particular umbrella. The last part contains some potential areas of research and development that are found to be lacking or understudied in the security software protocols of VANETs and that could lead to interesting results in the future.

11.2 System Model

We must first define the standard features of VANETs before discussing any theoretical or practical application of security in the networks. It is also important to understand the constraints shared by a majority of VANETs, be they physical or architectural. We will also define the constraint that security systems and broadcasted messages must fulfill in such a network.

11.2.1 Network Model

The nodes present in VANETs can be either a mobile vehicle or a Roadside Unit (RSU). Vehicles can be either private or public while RSUs can belong to governmental authorities or private service providers such as telecommunication providers but in some cases, a parked vehicle can be considered a RSU. The large scale of the VANETs is another main feature that creates the need to set them apart from other IoT networks and MANET. With countless nodes distributed on a broad scope, VANETs are most likely to become the largest actual application of a MANET. However, due to its application and cluster-based topography, communication in the network at large will be mainly local, thus allowing for the partition of the network and easing its scalability [7].

The greater majority of the network nodes consist of vehicles, thus the network dynamics are characterized by high speeds, quasi-permanent mobility and short connection times between members. One aspect of network dynamics that makes it slightly less chaotic is the predictable vehicle trajectories which are mostly well defined by the roads, which offers some advantages for the dissemination of messages and the reduction of randomness compared to most MANET while creating the disadvantages of an increase in potential breaches leading to disturbances in the provided services.

Another distinct and major advantage of VANETs compared to more traditional ad hoc networks is the nonnegligible computational and power resources the vehicles can easily provide. A standard "smart" vehicle in a VANET hosts hundreds or even thousands of microprocessors, an Event Data Recorder (EDR) that can be put to use for incident reconstruction, and a geolocational system that can indicate position and time.

The primary function of VANETs should be operational without dependence on heavy infrastructure presence nor the majority of other "smart" cars on the road as the two will keep coexisting for the following decades.

11.2.2 VANET Application

We have divided the potential uses of VANETs into two different categories depending on their level of importance from a security point of view. They have a vast range, and some are currently in use while others are more futuristic and are only discussed in theory.

The first one is the safety-related applications, with a concrete example being cooperative driving and collision avoidance. The defining characteristic of this type of application is its relevance to emergency situations where the presence of a service may prevent or help cope with life-threatening accidents. The security of this category, be it physical or electronic, is a must as the proper operations of any of these functions should be guaranteed even in the presence of disturbances, be it due to errors or attacks.

The other category includes the not urgent applications, including infotainment, location-dependent services, traffic optimization, and payment services. It cannot be denied that security is without a doubt a requirement in this category, especially when considering the cases where monetary transactions are involved. The examples of attacks and defenses presented further will mostly focus its attention on the security aspects of safety-related applications as they are often more concerning when looking specifically from the automotive domains perspective and because they usually raise the most challenging problems, possess harsher constraints, and carry greater risks. Some attention will still be given to other applications, especially concerning potential identity theft and economic damage.

11.2.3 Message Requirements

A security system developed for use in VANETs should by default be able to satisfy the following conditions and requirements for its safety messaging:

- **Authentication:** Reactions given by vehicles to events should be based on legitimate messages. Therefore proper protocols of authentication should be employed by the senders of these messages be it by public keys given out by the government or local network keys.
- **Data Plausibility:** The legitimacy of transmitted messages also includes the evaluation of their consistency with similar ones, as the legitimacy of the broadcaster can be assured while the contents of the message contain

erroneous data. The way that the plausibility is confirmed will firmly depend on the type of data transferred.

■ **Availability:** Even when we assume the existence of a robust communication channel, some attacks and malfunctions can weaken and bring down the network be it by sheer volume or finding flaws in the system. Therefore, it is paramount that availability of services should be also supported by alternative means. This can be a V2V or V2I solution, but a backup protocol should be in place.

■ **Time Critical:** Considering the very high speeds expected in a typical VANET, stringent constraints should be expected when dealing with time-sensitive data as it might not leave any room for mistakes and could have disastrous results if not met properly.

■ **Non-repudiation:** Individuals and vehicles causing accidents need to be reliably identified while the sender should not be able to pick and choose which message to broadcast or deny the transmission of certain messages. This leads us to the last point which might be the hardest to address in this hyper-connected world.

■ **Privacy:** People have become more aware and concerned about tracking and invasive surveillance technologies. The privacy of users and members against unauthorized observers has to be guaranteed. This is a chief concern as the development of VANETs will follow customer demand as car companies will most likely not innovate if it cannot sell its ideas.

11.2.4 Message Constraints

We mainly concern ourselves with safety applications and applications where identity theft leading to economic or social damage is a possibility. Using this perspective, we can classify the safety messages into four distinct classes by weighting their properties related to privacy and real-time constraints.

■ **Traffic information:** disseminates traffic conditions that might arise from natural or unnatural causes in a given region and thus affect public safety only indirectly; hence they are often not time dependent but require general privacy and legitimacy.

■ **General safety:** employed by public safety applications such as cooperative driving and collision avoidance and must satisfy strict time constraints. The fact that "smart" and "stupid" vehicles coexist in the system simultaneously means that identifying them in time will be essential for VANETs to work properly.

■ **Liability:** distinguished from the previous class as they are exchanged in a situation where liability plays a part such as speeding, collisions, accidents or unusual events. Before the broadcast, the extent of the liability held by the originator of the message should be determined and reveal his identity to the authorities should it prove necessary. This is also one of the problems in effective integration of VANETs and market penetration. Selling vehicles that might report their owner to the law is a tough sell. It is an ethical question that must be solved and answered as to when the liability messages should be broadcasted to the authorities.

■ **Identity-Economic messages:** are not counted as critical yet are becoming an integral part of VANET communications as automatic toll payment, and wireless payment options become more prolific. Malicious individuals might intercept payment information and use them without the knowledge of the user. Stolen identity, be it that of an individual or car, can create liability and security-related issues.

11.2.5 Trust and Privacy

The primary element in any security system is trust and privacy. This is particularly true and critical in VANETs due to the high liability expected from safety and security applications and consequently the members running them. With a significant number of independent nodes in the network and the presence of the human factor, it is without a doubt highly probable that misbehavior will happen. In a connected world, users are increasingly concerned about their privacy and drivers are by no means an exception. This is especially problematic as the lack of privacy and the potential tracking functionality inherent in VANETs may result in fines on the drivers, leading in turn to a mistrust by the users toward VANETs. Due to the previous point, we should likely assume a low amount of trust in members and service providers. Beside the two main members, the drivers and service providers, it is without a doubt certain that there will be, and should be, a considerable presence of governmental authorities in these networks.

11.3 Attacks in VANETs

As it is not realistic to envision and name all the potential attack avenues that can be mounted in the future on VANETs, we will strive to provide a general classification of both the possible forms taken by the perpetrators and the type of attack identified in the literature. We concern ourselves with VANETs; we will only consider and take into account the attacks that can be perpetrated against messages rather than directly on the vehicles, as the physical security of vehicles OBUs are outside of our scope and would require a whole other survey.

11.3.1 Attacker Classification

Understanding the nature of the attacker is important for classifying the types of attacks that VANETs might be subjected to. This classification is straightforward and is commonly accepted among researchers and engineers dealing with the subject.

■ **Insider/Outsider:** An authenticated member of the network who can broadcast and receive messages from other members is an insider. It has access to a certified public key and can more easily navigate the network protocols for mounting attacks. Meanwhile, the outsider is considered as a foreign object by the network members and as an intruder. Hence, it is severely limited in his interactions with security protocols.

■ **Malicious/Rational:** A malicious attacker is not in search of any kind of personal benefits from the attacks he perpetrates and aims to harm the members or the integrity of the network and create havoc. The main issue is the fact he could employ any means disregarding corresponding costs and consequences making him unpredictable and potentially extremely dangerous. On the contrary, an attacker that seeks personal profit is rational and will not overextend his resources for any intangible gain. Hence, this makes him more predictable regarding the attack means and the attack targets that he might choose to assault.

■ **Active/Passive:** Active attackers generate packets or signals, meanwhile a passive attacker contents himself with eavesdropping in on the wireless channel.

■ **Local/Extended:** An attacker can be limited in scope, even if he controls several nodes, which makes him local and limits his impact at large. The extended attacker can control several entities that are scattered across the network, thus allowing him to be active on a larger scope. The differentiation between local and extended is not easy to make and depends heavily on the size and range of the VANETs in question. This distinction can be especially important in tracking privacy-violating activities and potential suspects.

11.3.2 Attacks

The following examples are the types of attacks discussed in the literature. Some of them are simple to set up and might even be unintentional or just natural misbehaviors in the network rather than the result of a conscious effort to disrupt the active system. We should also assume that as the complexity of the attack method increases the skill of the perpetrator increases accordingly. Attacks that deal mainly with tampered hardware and OBU are not part of the scope of this review [8–16].

■ **Masquerade:** The member is actively masking its own identity to appear like another vehicle by using false identities, such as public keys. This technique is usually employed in conjunction with other types of attacks.

■ **False information:** Transmits erroneous information and data in the network, which might affect the behavior of other drivers. It can be both intentional and unintentional.

■ **Cheating with sensor information:** Attackers alter their perceived location and speed to escape liability. It can be used notably in the case of an accident. In the worst case, colluding attackers can clone each other, but this would require retrieving the security material, a subject outside the scope of this work.

- **Location Tracking:** The observer can monitor the trajectories of selected members and can use this information for a range of purposes, both malicious and mundane. It can also potentially leverage on the RSU or the vehicles who are around its main target. It is more likely than not malicious and requires some preparation, such as disseminating a virus in said network or some prior physical access.

- **Denial of Service (DoS):** An attacker can bring down VANETs, jam signals, or may even cause an accident by using malicious nodes to forge a significant number of bogus identities, such as IP addresses, with the final objective of disrupting the proper functioning of data and information transfer between two fast-moving members. An example would be jamming the communications channels. It is the most likely of basic attacks perpetrated by a malicious attacker. Preventing DoS attacks has been a research topic of computer sciences and an extensive amount of material is available [17–19] that could be applied to VANETs such as the IP-CHOCK scheme proposed in Verma et al. [20]. The IP-CHOCK approach is capable of locating suspicious nodes without requiring any kind of special hardware support or additional secret information exchange. Another approach is the Attacked Packet Detection Algorithm (APDA) presented in Roselin et al. [21] where it is employed to detect and identify the DoS attacks before engaging in resource-intensive verification time in an effort to increase security and minimize overhead delay.

- **GPS Spoofing:** GPS satellites, or their equivalent, maintain a locational table with the spatial locations and identifiers of vehicles in the network. Attackers may produce misleading and false readings in the positioning system with the purpose to deceive vehicles, leading them to assume that they are actually in a different spot. It is relatively easy to dupe any number of members as discussed in Tippenhauer et al. [22] with some restrictions. It is also quite possible to use a GPS satellite simulator to generate and broadcast signals that are stronger than those generated by the actual satellite system, leading the receivers to prefer it to the actual satellite.

- **Physical/Electronic Tunnel:** When a GPS signal disappears in a tunnel, it can be exploited by using the temporary loss of connection to the system to inject falsified data and positioning information. Once the vehicle leaves the tunnel, it will assume that this is his actual position and will tread as such before it receives an updated position from the satellite. A non-physical tunnel could be created using proper jamming which broadens the potential applications of such an attack.

- **Wormhole:** Traditionally this is accomplished by tunneling packets between two remote members of a network [23]. In a VANET, the perpetrator should control at least two nodes separate from each other and with a very high-speed connection between them to tunnel packets from one location to broadcast them in another. This could be accomplished with pre-established RSUs or by using mobile technologies such as 4G, or 5G in the near future. Wormholes allow the attacker to spread misleading but properly signed messages at the destination area. A way to protect a network from wormhole attacks is the TESLA with Instant Key disclosure (TIK) [23]. It is a method using packet leashes that calculates the divergence in the allowed theoretical travel distance and the actual travel distance of the transmitted packet to identify and spot any signs of tampering on the anomaly that might point to a potential attack. Another potential solution discussed in the literature is the AODV routing protocol where Hop-by-Hop Efficient Authentication Protocol (HEAP) [24] is a useful approach that allows us to notice wormhole attacks.

■ **Black Hole:** Data packets get lost while crossing through a black hole, in effect a member that has some nodes or no node that refuse to broadcast or forward data packets to the next hop. Preventing black hole attacks is generally achieved by making use of the redundant paths kept between the sender and the destination of the message, although this has the unfortunate side effect of adding to the network complexity. Another potential way to defend against black holes is the use of an information carrying sequence number in the message header. The receiver can then potentially figure out the absence of a packet in the case of any discrepancy or loss, identifying the situation as suspicious [15].

■ **Malware and Spam:** Attacks, like spam and viruses, can lead to severe disruptions in VANETs operations. They are typically the work of malicious insiders rather than outsiders who have access to OBUs of vehicles and RSUs when they are performing software updates. They can potentially result in an increase in transmission latency, which can be lessened by using centralized management. Proper maintenance of infrastructure and a centralized administration should be employed to prevent any such attack to the VANETs, the security of the OBUs of the mobile members are not in our scope.

■ **Timing:** Data transmission at the right time from one node to another is essential in achieving data security and integrity. An attack is labeled as Timing whenever a malicious vehicle receives a time-critical emergency message and does not forward it to their neighboring members on time by adding some additional timeslots to the original in an effort to create an artificial delay in transmission and reaction. Thus, neighboring vehicles to the attackers may receive the message outside the scope of the time constraint, rendering it moot.

■ **Man-in-the-Middle:** Malicious vehicles eavesdrop on the communication between vehicles and inject false information or distort messages between vehicles[14, 15]. Reasonable solutions are strong cryptography, secure authentication, and data integrity verifications.

- **Sybil:** The perpetrator creates multiple identities in an effort to simulate multiple nodes [25, 26]. Each of these nodes can transmit messages, even with multiple differing identities. Thus, other vehicles are led to think that there are more vehicles in the network than there actually is. This type of attack is potentially extremely hazardous in the constraint heavy VANETs since a vehicle can also potentially claim to be in a different spot at the same time causing substantial security risks and causing chaos in the network. They are traditionally detected through the use of resource testing [8, 12], but this approach assumes that all entities are bound to be limited to some resources. Computational PUZZLES are used in [8] to assess computational resource usages of each node. However, this technique cannot be said to be appropriate for use in Sybil attack detection in VANETs [12] as an attacker node can have potentially more computational power than an ordinary node encountered in other MANET and IoT systems. Instead of computational testing, the use of radio resource testing [12] can be used to identify the attack as the attacker node will most likely use more broadcast resources than standard nodes as it will broadcast like multiple members simultaneously. The use of public key cryptography [27] can prevent and eliminate the risk of Sybil attacks where the vehicles are authenticated using the public keys distributed by authorities. Key revocation can be another possible approach that may reduce the influence and impact of Sibyl attacks identified in wireless sensor networks [28, 29] using a predefined model of propagation. Using this model we can measure the true distance of a node through the Received Signal Strength Indicator (RSSI) method, where the differences of the signal strength of the transmitted and received messages are compared while signals are matched with the nodes claimed spatial location. If the claimed spot by the node does not correspond to the defined distance limit set from the evaluated location for that model of propagation, this node is a potential Sybil attacker and should be treated as such.
- **Illusion Attack:** This is when an attacker broadcasts the traffic warning messages that do not correspond to the current road condition, producing an illusion to the members in their neighborhood. The propagation of the phantasm mainly depends on the drivers' responses, which can lead to car accidents, traffic jams, and a decrease in the overall health and performance of the VANET. Classic message control approaches cannot protect networks against illusion attacks due to the explicit control the adversary exerts on the sensors of its vehicle, misleading its inputs in an effort to create and broadcast false information. The Plausibility Validation Network (PVN), proposed in Lo and Hsiao-Chein [11], is a potential model to secure VANETs against this type of assault. PVN collects raw sensor's data and verifies whether the collected data seems realistic and plausible. It requires two distinct types of

inputs: data transmitted from antennas and data sensed by sensors. They are then categorized by employing a header placed on the transmitted data. PVN possesses a rule database and checking module for transmitting data which allows it to check the authenticity of the inputs and take the necessary actions accordingly if found to be illusionary. A transmission is considered trustworthy if it can pass all verifications processes successfully. If it fails, it is treated as an illusion and dropped by default. PVN can potentially cooperate with different types of cryptographic methods to be able to defend against more various forms of attacks.

■ **Purposeful/Intentional Attack:** The attacks perpetrated by insiders are problematic to prevent and defend against properly as they are already an authenticated member of the network and most likely trusted to perform V2V and V2I communications with their neighbors, making them difficult to catch. In VANETs, where a disturbance might have harsh consequences, it is vital to defend against any such behavior. Misbehaving nodes can be made to deny forwarding messages that it receives from another member, purposefully misinterpret or modify messages, misuse the available bandwidth, or inject bogus information. The technique that has been proposed in Pires et al. [30] can be used to defend against misbehavior in V2I and V2V communications. It considers the authenticity of anonymous communications in an effort to prevent potential misbehavior while maintaining the privacy of its members. The threshold authentication technique is used where a threshold value is set up to identify strange, unusual, or misbehaving nodes a predetermined number of times. If there are any repeated offenses or result over the threshold value, it will trace the misbehaving node's credential.

■ **Impersonation Attack:** During V2V communications, a member can broadcast the security messages as if it was the origin to other vehicles which can potentially have an impact on the behavior of the traffic control system and the other members of the VANET. This a case of an impersonation attack, when a malicious vehicle owner transmits a message on behalf of another member to cause accidents, traffic jams, create chaos, or other security attacks while masquerading. The SPECS scheme, proposed in Chim et al. [31] and detailed in Chim et al. [32], ensures the privacy and security of V2V communications by detecting the impersonation attacks. This is accomplished with the use of an Identity-based batch Verification protocol which suffers from this type of attack and cannot fulfill its privacy requirements, leading to the discovery of the misbehaving member. To protect the identity of each vehicle, it uses a pseudo-identification code and a shared secret key between the RSU and vehicles present in the network. This type of attack is the reason the distribution of identity keys by the authorities is paramount for identifying the origin of the broadcasted messages.

11.4 Defenses

Existing security and privacy schemes for VANETs can be overall classified into four broad approaches. They all contain different schemes and protocols to ensure the safety of VANETs. The presented protocols main objective is protecting the messages rather than safeguarding the systems themselves against an attacker once it has infiltrated or falsified the transmitted broadcasts.

11.4.1 Public Key

A node in the VANET is given a pair of keys, public and secret. A proper Public Key Infrastructure (PKI) should, in theory, be able to efficiently handle the management of keys to provide security. Usually, a scheme that contains the use of a PKI is generally proposed whenever a vehicle has two specific hardware units: Tamper Proof Hardware (TPH) and EDR to perform cryptographic processes and record all the events.

Hesham et al. [33] propose a protocol using a dynamic method of key distribution that can handle key management which does not have a need to store a large number of keys in its memory for employing PKI, thus reducing the amount of TPH needed. Using this approach as a basis, member-specific unique information such as Electronic License Plate (ELP) or chassis number that forms the Vehicle Authentication Code (VAC) can be employed as a hidden key shared by a certificate authority (CA) and a member. CAs are responsible for issuing, renewing, revoking, and distributing public keys [34]. The use of this protocol grants a strong resistance to DoS, Sybil, and Man-in-the-Middle attacks since it makes use of ELP and a unique VAC encrypted secret key to protect against masquerade attempts.

Gazdar et al. [35] propose an effective dynamic cluster-based architecture for a PKI defense model in VANETs employing a trust model where a value ranging from 0 to 1 is attributed to them. Based on this value, members can have four different roles being CA, Member Node (MN), Registration Authority (RA), and Gateway (GW). RA and CA which are MN that have a trust value that is equal to 1 can issue certificates to the members in the cluster while protecting the CA against an attacker by avoiding any potential direct contact and exchange between the CA and an untrusted vehicle. A GW is employed for inter-cluster communications. Nodes, including MNs, have to demonstrate good cooperation and comportment to increase their values of trust. In this architecture, a hierarchical monitoring scheme is employed to observe the comportment of its members, where a member with higher trust value monitors the vehicles with lower trust values. The PKI based scheme of Efficient Certificate Management Scheme (ECMV) [36], provides an effective certificate management control between different authorities, granting the OBU with the capability of updating its certificates at any instant regardless of its location. Even if an adversary manages to infiltrate the network, an ECMV has a solid certificate revocation protocol to follow to remove the intruder. This scheme has proven to reduce the complexity of managing certificates to a large extent and can be very effective in making PKIs scalable and more secure, which in turn makes it ideal for IoT and VANETs in general.

11.4.2 Hybrid and Symmetric

In these types of schemes, nodes start to communicate only after they agree to share a secret key that will be employed for securing communications. Most of the current security schemes available to VANETs are dependent on either symmetric or public keys to encrypt communications. A potential hybrid system that employs both public and symmetric keys has been proposed for securing VANETs [14], using the two different types of communications, being group and pair-wise. During the group communication, more than two nodes can communicate whereas the pair-wise communication is employed when two members need to communicate with one another. The hybrid approach employs symmetric keys for pair-wise transmissions in an effort to avoid the potential overhead that happens when using the key pair. It should be noted that symmetric keys must not be employed during an authentication process as it will prevent non-repudiation which is a main constraint of VANET messages [15].

11.4.3 Certificate Revocation

In the broader scope of PKI methodology used to provide security to VANETs, there exists a major category called certificate revocation [34]. Certificate revocation is usually applied and enforced by the CA in two ways: centralized and decentralized. The first option employs a centralized authority which is only responsible for making the revocation decisions whereas, in the second option, a group of neighboring members of the vehicle to be revoked make the decision. This scheme is more often than not centralized and requires pervasive infrastructure and thus cannot be said to be efficient since RSUs need to send the certificate revocation list to OBUs while the deployment cost is relatively high compared to most other PKI methods. This RSU dependence could be alleviated in the near future with the use of 5G technologies and cloud computing and sharing. For now, a couple of alternative approaches and implementation have been put forward such as the Distributed Revocation Protocol, the Revocation Protocol of Tamper Proof device, and the Revocation Protocol using Compressed Revocation Lists. Another newly proposed scheme is the RSU aided Certificate Revocation, where a trusted third party grants different secret keys for each individual RSU so that they can additionally sign the messages transmitted in its range. Once a member's certificate is proven to be invalid, CAs broadcast messages to the RSUs which in turn transmits orders to all vehicles in range to revoke that particular vehicle's certificate and to stop communication with it, effectively isolating the problematic member from the network.

11.4.4 ID-Based Cryptography

PKI and symmetric key cryptography are not always the most optimal schemes to provide security and robustness to VANETs since they are most of the time infrastructure dependent and lacking additional layers of protection. An alternative, the ID-based cryptography, that contains the best features of the traditionally employed security schemes and protocols are being explored by the scientific community. ID-based cryptography reduces the required computational resource cost in the ID-based Signature (IBS) processes and is preferable for using the ID-based Online/Offline Signature (IBOOS) scheme for authentication in VANETs. IBOOS has been shown to increase efficiency by dividing the signing process into two phases, online and offline, in which the verification has been shown to be more efficient than that of IBS.

Lu et al. [37] propose an ID-based framework that makes use of both IBS and IBOOS for authentification. It should employ self-defined pseudonyms instead of any kind of real-world IDs to protect the member's privacy and confidentiality. It has proven to be efficient in term of communication overhead, storage, and processing time. This is achieved by preloading a pool of IDs for the overall regional RSUs in each vehicle beforehand, which is often very small in size and is not expected to change as frequently compared to some of the other approaches that make use of pre-stored IDs of all the potential RSUs. This is done by using IBS for V2R authentications while IBOOS is employed for V2V authentications. Evaluation results have shown that this scheme manages to efficiently preserve the privacy of VANETs.

Pan et al. [38] have put forward a model to quantify the locational privacy by applying a simple scheme called Random Changing Pseudonyms. Each member changes its attributed pseudonym after an arbitrary point in time and space. It should be noted that it is very important to provide unlinkability for the two successive pseudonyms of any member. If this constraint is not fulfilled an intruder might potentially be able to locate the tracked member by mapping the link between the changing pseudonyms. The probability of the pseudonyms unlinkability is directly affected by the efficiency of the various pseudonym-shifting schemes in use to protect locational privacy.

11.5 Research Opportunities

The quick development of VANETs has created a great number of open research possibilities, starting with securing them against inside and outside threats. Any research aiming to improve the robustness and reliability of defensive schemes is always open to more in-depth work. The development of smart key attribution be it governmental or local, and not linked to IP or MAC addresses, is a necessity as the ever-expanding scalability of the network will most likely not support current identification schemes. The ethical issues should also be discussed in depth and proper liability thresholds should be established to encourage the dissemination and market penetration of VANETs. The matter of fact use of the internet and remote transaction in our daily lives, such as paying a highway toll wirelessly, has made it a concern when dealing with identity theft. This makes the development and design of secured communication protocols for VANETs with an emphasis on protecting user profiles, private data and information from malicious vehicles and infrastructure has to be given a higher priority than before where it was relegated as a secondary concern. A more general topic would be the integration of 5G communication technologies in VANETs and the potential for cloud computing and virtual infrastructure for certificate revocation defense protocols. Another pertinent avenue of research is the scalability of the proposed methods as a defense method which is not designed and stress-tested carefully for a high-density and high-traffic situation will most likely fail due to not being able to fulfill constraint or crumble due to errors. Comparison between the efficiency of the four main defense schemes could yield interesting results concerning optimization of VANET security protocols. The analytical models to accurately quantify the effectiveness of pseudonym-changing schemes to reinforce unlinkability is a major research topic to provide security in VANETs using ID-Based Cryptography as the efficiency of the scheme can depend mainly on this property. Finally, a potential approach that defines trust, taking into account the expected behavior of that vehicle compared to its previous broadcasts and standard movement patterns might be a future defensive scheme.

11.6 Conclusion

The propagation of VANETs will most likely continue and it will become a seamless part of our lives in the future. For this to happen, the major concerns concerning security being privacy for the users, and robustness for the health of the network, should be guaranteed and the constant development of new protocols and schemes is a good indicator of how much the scientific community is aware of this reality. The topics of research concerning software development and technical subjects abound as demonstrated in the relevant passage and it should be noted that a lot of such concerns also exist when we take a look at the more physical part of the equation concerning OBUs and how to effectively make them tamper proof. The question of liability and government control and management is a whole other subject that requires the involvement of policymakers and the consensus of the public to answer and establish where responsibility begins and ends. With the advent of potentially self-driving vehicles, establishing an infrastructure, both physical and electronic, where both human controlled and automated members interact harmoniously will become an important factor for the proper implementation of automated vehicles and VANETs.

References

1. Gerla, Mario, et al. "Internet of Vehicles: From Intelligent Grid to Autonomous Cars and Vehicular Clouds," *Internet of Things (WF-IoT), 2014 IEEE World Forum on. IEEE*, 2014.
2. Gubbi, Jayavardhana, et al. "Internet of Things (IoT): A Vision, Architectural Elements, and Future Directions," *Future Generation Computer Systems* 29.7, (2013): 1645–1660.
3. Andrews, Jeffrey G., et al. "What Will 5G Be?", *IEEE Journal on Selected Areas in Communications* 32.6, (2014): 1065–1082.
4. Zeadally, Sherali, et al. "Vehicular ad hoc Networks (VANETS): Status, Results, and Challenges," *Telecommunication Systems* 50.4, (2012): 217–241.
5. Alam, Kazi Masudul, Mukesh Saini, and Abdulmotaleb El Saddik. "Toward Social Internet of Vehicles: Concept, Architecture, and Applications". *IEEE Access* 3 (2015): 343–357.
6. Sun, Wei. "Internet of Vehicles," *Advances in Media Technology*, 47, 2013.
7. Fangchun, Yang, et al. "An Overview of Internet of Vehicles," *China Communications* vol.11, no.10 (2014): 1–15.
8. Douceur, John, R. "The Sybil Attack," *International Workshop on Peer-to-Peer Systems*. Berlin, Heidelberg: Springer, 2002.
9. Guette, Gilles, and Bertrand Ducourthial, "On the Sybil Attack Detection in VANET," *IEEE International Conference on Mobile Adhoc and Sensor Systems (MASS'07)*. IEEE, 2007.
10. Raya, Maxim, and Jean-Pierre Hubaux, "Securing Vehicular ad hoc Networks,". *Journal of Computer Security* , 15.1 (2007): 39–68.
11. Lo, Nai-Wei, and Hsiao-Chien Tsai, "Illusion Attack on Vanet Applications—a Message Plausibility Problem," *Globecom Workshops, IEEE*. 2007, 2007.
12. J. Newsome, E. Shi, D. Song, and A. Perrig, "The Sybil Attack in Sensor Networks: Analysis & Defenses". *Proceedings of the 3rd International Symposium on Information Processing in Sensor Networks*. ACM, 2004 Berkeley, CA, USA.
13. Sumra, Irshad Ahmed, Iftikhar Ahmad, and Halabi Hasbullah, "Behavior of Attacker and Some New Possible Attacks in Vehicular Ad hoc Network (VANET)," *3rd International Congress on Ultra-Modern Telecommunications and Control Systems and Workshops (ICUMT'11)*, IEEE, 2011 Budapest, Hungary.
14. F. Al-Turjman, and S. Alturjman, "Context-sensitive Access in Industrial Internet of Things (IIoT) Healthcare Applications," *IEEE Transactions on Industrial Informatics*, vol. 14, no. 6, pp. 2736–2744, 2018.
15. Al Hasan, Ahmed Shoeb, Md Shohrab Hossain, and Mohammed Atiquzzaman. "Security Threats in Vehicular ad hoc Networks". *International Conference on Advances in Computing, Communications* and *Informatics (ICACCI'16)* IEEE, 2016 Jaipur, India.
16. La Vinh, Hoa, and Ana Rosa Cavalli, "Security Attacks and Solutions in Vehicular ad hoc Networks: A Survey," *International Journal on AdHoc Networking Systems (IJANS)* 4.2 (2014): 1–20.
17. Agah, Afrand, and Sajal K. Das, "Preventing DoS Attacks in Wireless Sensor Networks: A Repeated Game Theory Approach," *IJ Network Security* 5.2 (2007): 145–153.

18. Tagra, Deepak, Musfiq Rahman, and Srinivas Sampalli. "Technique for Preventing DoS Attacks on RFID Systems," *International Conference on Software, Telecommunications and Computer Networks (SoftCOM'10)*, IEEE, 2010 Dubrovnik, Croatia.

19. Mohi, Maryam, Ali Movaghar, and Pooya Moradian Zadeh, "A Bayesian Game Approach for Preventing DoS Attacks in Wireless Sensor Networks," *International Conference on Communications and Mobile Computing, 2009. CMC'09. WRI*, Vol. 3. IEEE, 2009.

20. Verma, Karan, Halabi Hasbullah, and Ashok Kumar. "Prevention of DoS attacks in VANET," *Wireless Personal Communications*, 73.1, (2013): 95–126.

21. RoselinMary, S., M. Maheshwari, and M. Thamaraiselvan. "Early Detection of DOS Attacks in VANET using Attacked Packet Detection Algorithm (APDA)," *International Conference on Information Communication and Embedded Systems (ICICES'13)*, IEEE, 2013 Chennai, India.

22. Tippenhauer, Nils Ole, et al. "On the Requirements for Successful GPS Spoofing Attacks," *Proceedings of the 18th ACM Conference on Computer and Communications Security*, ACM, 2011 Chicago, Illinois.

23. S. Alabady, and F. Al-Turjman, "Low Complexity Parity Check Code for Futuristic Wireless Networks Applications," *IEEE Access Journal*. vol. 6,no. 1, pp. 18398–18407, 2018.

24. Safi, Seyed Mohammad, Ali Movaghar, and Misagh Mohammadizadeh., "A Novel Approach for Avoiding Wormhole Attacks in VANET," *Second International Workshop on Computer Science and Engineering, (WCSE'09)*, vol. 2. IEEE, 2009 Qingdao, China.

25. F. Al-Turjman, "Fog-based Caching in Software-Defined Information-Centric Networks," *Elsevier Computers & Electrical Engineering Journal*. vol. 69, no. 1, pp. 54 –67, 2018.

26. Park, Soyoung, et al. "Defense against Sybil Attack in Vehicular ad hoc Network Based on Roadside Unit Support," *Military Communications Conference, 2009 (MILCOM'09)*. IEEE, 2009 Boston, Massachusetts.

27. Raya, Maxim, Panos Papadimitratos, and Jean-Pierre Hubaux. "Securing Vehicular Communications," *IEEE Wireless Communications* 13.5 (2006).

28. Xiao, Bin, Bo Yu, and Chuanshan Gao. "Detection and Localization of Sybil Nodes in VANETs," *Proceedings of the 2006 Workshop on Dependability Issues in Wireless Ad Hoc Networks and Sensor Networks*. ACM, 2006 Los Angeles, California.

29. Sun, Jinyuan, and Yuguang Fang. "A Defense Technique against Misbehavior in VANETs Based on Threshold Authentication," *Military Communications Conference, 2008. MILCOM' 08)*. IEEE, 2008 San Diego, California.

30. F. Al-Turjman, and S. Alturjman, "Confidential Smart-Sensing Framework in the IoT Era," *The Springer Journal of Supercomputing*, vol. 74, no. 10, pp. 5187–5198, 2018.

31. F. Al-Turjman, "5G-enabled Devices and Smart-Spaces in Social-IoT: An Overview," *Elsevier Future Generation Computer Systems*, 2017. DOI: 10.1016/j.future.2017.11.035

32. Chim, Tat Wing et al. "SPECS: Secure and Privacy Enhancing Communications Schemes for VANETs," *Ad Hoc Networks*, 9.2 (2011): 189–203.

33. Hesham, Ahmed, Ayman Abdel-Hamid, and Mohamad Abou El-Nasr, "A Dynamic Key Distribution Protocol for PKI-based VANETs," *Wireless Days (WD), 2011 IFIP*. IEEE, 2011.

34. F. Al-Turjman, "Information-Centric Framework for the Internet of Things (IoT): Traffic Modelling & Optimization," *Elsevier Future Generation Computer Systems,* vol. 80, no. 1, pp. 63–75, 2017.

35. F. Al-Turjman, "Cognitive Caching for the Future Sensors in Fog Networking," *Elsevier Pervasive and Mobile Computing*, vol. 42, pp. 317–334, 2017.

36. Wasef, Albert, Yixin Jiang, and Xuemin Shen. "ECMV: Efficient Certificate Management Scheme for Vehicular Networks," *Global Telecommunications Conference, IEEE GLOBECOM 2008*. IEEE, 2008 New Orleans, Louisiana.

37. F. Al-Turjman, Y. K. Ever, E. Ever, H. Nguyen, D. Deebak, "Seamless Key Agreement Framework for Mobile-Sink in IoT based Cloud-centric Secure Public Safety Networks," *IEEE Access*, vol. 5, no. 1, pp. 24617–24631, 2017.

38. F. Al-Turjman, "Mobile Couriers' Selection for the Smart-grid in Smart cities' Pervasive Sensing,"*Elsevier Future Generation Computer Systems*, vol. 82, no. 1, pp. 327–341, 2018.

Index